Philosophy of Physics

A new introduction

Online at: https://doi.org/10.1088/978-0-7503-2636-0

Philosophy of Physics

A new introduction

Robert P Crease
Department of Philosophy, Stony Brook University, New York, USA

IOP Publishing, Bristol, UK

ISBN   978-0-7503-2636-0 (ebook)
ISBN   978-0-7503-2634-6 (print)
ISBN   978-0-7503-2637-7 (myPrint)
ISBN   978-0-7503-2635-3 (mobi)

DOI   10.1088/978-0-7503-2636-0

Version: 20230301

IOP ebooks

British Library Cataloguing-in-Publication Data: A catalogue record for this book is available from the British Library.

Published by IOP Publishing, wholly owned by The Institute of Physics, London

IOP Publishing, No.2 The Distillery, Glassfields, Avon Street, Bristol, BS2 0GR, UK

US Office: IOP Publishing, Inc., 190 North Independence Mall West, Suite 601, Philadelphia, PA 19106, USA

Contents

Foreword

Like many other philosophers, I am constantly confronted with the challenge of explaining the meaning and value of philosophy of science to skeptical practitioners. Also like other philosophers, I have been confronted with the challenge of explaining certain approaches to philosophy of science to skeptical colleagues with who adopt different ones. Philosophers of science can talk, not just past scientists, but also past each other. To overcome this, I finally realized that I'd have to begin by explaining the different approaches. This book is the result.

Preface

Scientists are sometimes stymied by puzzles created, but not solved, by their own activities. What does quantum mechanics mean? What makes string theory scientific? Is there a fundamental branch of science? Are atoms 'parts' or 'moments' of molecules? How to define 'species', 'race', 'disease', 'biodiversity', 'element', 'discovery', 'theory', and other terms?

Left to their own devices, researchers can either ignore such puzzles or sweep them under the rug with rough and ready answers. But when the usual scientific theorizing and data-collecting is applied, the answers can contain ambiguities, contradictions, mysteries, or implausible verdicts. If you want to find philosophical issues in science, look here.

'Philosophy', like 'science', may refer to several different things. It may refer to a group of topics, such as beauty, truth, the good; to a set of discoveries; or to ways of inquiry. This book is about the latter. What are the ways that philosophers inquire, what happens when that inquiry is directed to science, and how can philosophy help resolve questions such as the above?

Two kinds of challenges

Being a scientist means inheriting a set of concepts, practices, and ways of thinking. Generally this inheritance works smoothly, but puzzles may arise when scientists find that their knowledge or expectations collide. When such puzzles are resolvable with more research they are *scientific* challenges; otherwise they may involve *philosophical* challenges.

Most scientific research does not generate philosophical challenges. Either that, or practitioners are confident that more experimental and theoretical work will clear things up—say, the fine details of superconductivity or of protein folding. Other areas of research are bedeviled by philosophical issues. No amount of physics seems able to solve the problem of how to interpret quantum mechanics in a satisfactory way—though one might say 'Not yet' and await future research.

The conviction that scientific inquiry, via the usual theorizing and data-collecting, will be able to resolve all such questions, and that it provides the only reliable way to settle any important question, is called *scientism*. But scientism is not a part of science; it is a story *about* science, one that claims to purge it of philosophical issues. This creates a set of problems and dangers, to be discussed in this book.

Philosophy has plenty to deal with. It inquires into the origin of scientism, into puzzles arising from collisions between features of science that science cannot repair, and into how scientific model-building springs from and illuminates our experience.

Coherence

Scientists and philosophers demand different kinds of coherence. Scientists demand the coherence of nature. If the energy of the $2s_{1/2}$ electron orbital in atomic hydrogen is not where quantum electrodynamics says it should be, if two ways of measuring the fine structure constant reach different values, or if the way tau and theta particles decay means that they can and cannot be the same particle, physicists must find explanations. Otherwise it threatens the foundations of the scientific enterprise itself.

Philosophers demand a different kind of coherence, the coherence of experience. What happens when we desire, savor, appreciate, remember, imagine, hallucinate, wonder, inquire, and are curious? How do these experiences relate to the theoretical, knowledge-seeking attitude of scientists in a laboratory? Who are we, we humans, that we can have all these experiences?

In telling the story that the usual theorizing and data-collecting is able to answer all those questions, scientism maintains that the first kind of coherence effectively absorbs the second, rendering philosophical activity unnecessary—something that mystifies rather than clarifies.

According to scientism, for instance, humans don't *really* see sunsets and sunrises but the earth spinning, we don't observe rainbows but reflections and refractions, what looks to be stars in the sky are only left-over glimmers of faraway objects that ceased to exist billions of years ago, the human body is 60% water, and so on.

No doubt such storytelling, which delegitimizes human experience, is meant to foster appreciation for science. But it can have the opposite effect.

Two tables

Almost a century ago, the English astronomer Arthur Eddington began his book *The Nature of the Physical World* with a famous image involving two tables. One is a familiar commonplace object of experience that is extended, colored, and motionless. Another table is the one as dissected by science, consisting mainly of tiny electrical charges swarming in empty space.

Physicists, Eddington said, may claim that the scientific table is 'the only one which is really there'—but they will 'never succeed in exorcizing' the table of ordinary experience. Eddington admitted that he was unsure how the two tables are related, saying that the question lies outside the scope of physics. Many of his colleagues, though, are sure, and say the scientific table is real and the ordinary one is just in our heads. There is no need to relate them.

It's fortunate that Eddington didn't ditch his ordinary, experienceable table. For even as he was writing his book, Werner Heisenberg and others were developing quantum mechanics, forcing Eddington to revise some of his manuscript. Evidently, even 'scientific' tables can be defective and swapped out for newer models.

Eddington was wise to say that ordinary experience can't be exorcised, for it's the precondition for any knowledge at all. Even when we've been mislead—say, in optical illusions where straight lines appear bowed—it's not that something outside me drops into a hole to replace the knowledge that the lines are angled by the fact that they are straight. I have to be curious, play around, and recognize that what I find integrates with what I already know, in an ongoing, temporally spread-out experiential process. It's not that somebody tells me, 'Don't believe your eyes, the lines *are* straight!' That's an opinion that I may or may not believe, and anyway I still *see* the lines as bowed. It's by using a ruler and maybe some explanation that I realize how I see the lines as bowed. I don't suspend my experience in realizing that something's an illusion, but enrich it.

Towards the end of his life, the German philosopher Edmund Husserl began an essay that he called 'The original Ark, Earth, does not move'. The editors of his posthumous works thought this sounded too absurd in a post-Copernican world, and retitled it 'Foundational Investigations of the Phenomenological Origin of the Spatiality of Nature'. But the point of the essay was exact. Experiencing a non-moving background environment is not a mistake, Husserl argued, but a precondition for humans to be able to develop a sense of movement and to model oneself as being on a moving Earth.

Philosophy of science involves examining the relation between such background environments and the scientific activities that arise in and thanks to them.

Scientific and philosophical stances

Near the beginning of his *Lectures on Physics* Richard Feynman captured his conception of physics in a dramatic image:

> We can imagine that this complicated array of moving things which constitutes 'the world' is something like a great chess game being played by the gods, and we are observers of the game. We do not know what the rules of the game are; all we are allowed to do is to *watch* the playing. Of course, if we watch long enough, we may eventually catch on to a few of the rules. *The rules of the game* are what we mean by *fundamental physics* (Feynman *et al* 1963, section 2-1).

To continue Feynman's metaphor, if scientists watch the playing to figure out the rules of the game, philosophers watch how scientists watch the game. Philosophers can ask many different questions about the process. How do scientists judge that they are watching a game, and decide what a 'rule of the game' is? How do they find those rules, make mistakes about them, know when they are mistaken, and correct those mistakes? What built-in assumptions about such things as material, motion, identity, time, and space have the scientists made in dividing up the game into pieces, the board, and the 'gods' (the 'movers'). What is involved in scientists narrowing their experience to the theoretical task of figuring out the game at all? What philosophers of science investigate is not so much the objects that scientists investigate but how scientists investigate them.

Philosophers of science and scientists investigate different things and are standing in different places. The philosophical and scientific stances involve different approaches, methods, concepts, standards, interests, vocabularies, and literatures. The scientific stance objectifies what is under investigation, treating it as something independent of mental activity, while the philosophical stance is aware of, informed by, and cannot leave out the human world to which philosophers, scientists, and the subject of inquiry all belong. A first task of philosophy of science is to convey the difference between the naturalist and philosophical stances.

The fact that philosophers and scientists have different stances means that it can be difficult for the latter to appreciate the former. It can lead scientists to assume that any difficulty they have in grasping what philosophers say must be due to the failure of philosophers to properly understand the subject-matter. If one makes that assumption, the natural response is either to break down the subject for the philosophers as if explaining it to novices, or to dismiss philosophers and philosophy altogether as irrelevant, wrongheaded, or obsolete.

This has been the case since the dawn of modern science. In 1604 Galileo published a pamphlet ridiculing philosophers who could not figure out whether a new star that had appeared the previous year was near Earth, as Aristotelians insisted, or at a remote distance like the other stars, as per astronomers. It is easy to find the answer, Galileo knew, if you understand the mathematics of parallax, or how from the perspective of a moving observer an object changes position against the background. Galileo had fun skewering philosophers who arrogantly regarded themselves wise in the ways of the world but were clueless about how to use a simple tool that might help them find out.

Galileo's fictional philosophers were indeed arrogant if they tried to use a scientific tool such as parallax to address philosophical issues. But Galileo was displaying his own arrogance if he assumed that those philosophers were essentially looking for the same thing that the scientists were. That assumption has persisted ever since. As Nobel laureate Steven Weinberg wrote in a 1977 *American Scientist* article:

> I have considerable difficulty in understanding the philosophical content that many people seem to find in discoveries in physics. It is true, of course, that many of the subjects of physics—space and time, causality, ultimate particles—have been the concern of philosophers since the earliest times. But in my view, when physicists make discoveries in these areas, they do not so much confirm or refute the speculations of philosophers as show that philosophers were out of their jurisdiction in speculating about these phenomena (Weinberg 1977, p 175).

Or, as physicist Murray Gell-Mann commented a few years later,

> In my opinion, worrying about philosophy is often bad for the fundamental theorist. It muddies the waters and obscures his principal task, which is to find a coherent structure that works. Moreover, a philosophical bias may easily cause him to reject a good idea (quoted in Crease 1983, p 9).

But philosophy is not a set of speculations, positions, or set of theories, and it definitely does not aim at solving scientific problems.

Philosophy can often be difficult to explain to scientists. Its language, like the discourse of science itself, frequently takes a narrow focus and is preoccupied with special topics and technical issues whose value understandably may not be obvious to outsiders. Scientists, meanwhile, may adopt the attitude that only the measurable is meaningful. That makes the phenomenal, qualitative world that philosophers typically address—the human world in which people actually *live*—appear less tangible, concrete, and interesting than the grandeur of things like Newtonian physics or the intricate beauty of quantum mechanics. Yet science and its 'magnificent structures' (as Francis Bacon put it) arise out of that human world and would be impossible without it.

Science without philosophy

The scientists' pursuit of the coherence of nature assumes a sharp boundary between that domain and ordinary experience. Adherents of scientism take that as a natural, and hard, boundary. They make a sharp distinction between experience and what scientists say that experience is of. The motive is educational and benign—to encourage an appreciation for the wonders of science.

But that sharp distinction is a problem. It encourages in outsiders the sense that there is a class of elite influencers (i.e. scientists) who think that only they possess the truth and that the rest of the population (i.e. non-scientists) is confused and misled, which makes it easier for people to ignore the claims of the elites as untrustworthy. Maybe sea levels are not rising, glaciers not melting. Maybe vaccines cause autism, and rising amounts of human-produced CO_2 in the atmosphere—if that's even true —is not responsible for global warming. Who knows? Why trust those influencers? Scientism can foster backlash—skepticism that only scientists have a true grip on the world and that the rest of us ordinary humans have only illusion or ignorance.

The philosophers' pursuit of the coherence of various types of human experience does not contain such a boundary, for it includes how the theoretical attitude arises and is related to other ways of experiencing. Education is not 'having the right information', given to us by experts who have mined the information from a special place to which only they have access. One task of philosophy of science is to develop a robust account of how scientific conclusions are generated from and only within human experience. That's not in the scope of science. It doesn't take a whole lot more humanities research to understand this, just a greater appreciation for it.

Philosophy of science is as alive, relevant, and as full of active questions as science itself. It is easiest to begin to present what it does, to physicists who may be interested in philosophy but know little about it, by outlining the aims and approaches of philosophy of science using the metaphor of science as a workshop, and then to exhibit the different approaches that can be taken to understanding that workshop and what goes on in it. The aim of this book is not to contribute to any one of these approaches, but to show how they fit together in the philosophy of science.

References

Crease R 1983 Good philosophy and good physics *Threepenny Rev.* Fall 9

Feynman R P, Robert B L and Sands M 1963 *The Feynman Lectures on Physics* (Reading, MA: Addison-Wesley)

Weinberg S 1977 The forces of nature: gauge field theories offer the prospect of a unified view of the four kinds of natural force—the gravitational and electromagnetic, and the weak and the strong *Am. Sci.* **65** 13–29

Acknowledgments

I owe many thanks to John Navas, the senior commissioning manager for the Institute of Physics (IOP) Publishing, who invited me to write a book explaining what the philosophy of physics was all about. He also encouraged me to use passages that I've written over the years—I can't believe that it's been over 22 of them—for my columns in *Physics World*, called 'Critical Point'. *Physics World* is the flagship of IOP publications, and a wide-ranging, entertaining, and, well, worldly magazine. Its editor, Matin Durrani, allowed me to explore a broad range of topics in my columns and edited them superbly, often challenging me when I lapsed into jargon or wandered too far from the point. Material from many of these columns are incorporated here. Thanks to graduate student and now Dr. Baxter Jephcott, who read the entire manuscript. I am also especially indebted to Jennifer Carter and Gino Elia, members of my philosophy of science reading group. Elia read an entire draft of the manuscript, and I am in awe of his insight into what I missed and was trying to say, in his boldness in being challenging and critical, and in his ability to articulate many points, which I incorporated, better than I did. In his comments in the right column of the marked-up manuscript he developed a character, the 'cynical physicist', whose wry, imaginative, and astute remarks I had to confront, for I knew that if I couldn't convince this character, I was not writing and even thinking well enough. Robert C Scharff, my philosophical conscience and mentor, also read many of these passages. His penchant for making extensive comments in track changes has given rise to a new adjective, 'scharffing', among those who correspond with him.

I struggled mightily and long with this manuscript, and was on the verge of giving up when I realized that the first step had to be to indicate the different approaches of philosophers to inquiring into what it's all about. Unless one understands and appreciates the value of these different approaches one cannot understand and appreciate philosophy of science. I hope that I have done enough to convert the cynical physicist, and to survive scharffing, to show the first steps as to how this is done.

Author biography

Robert P Crease

Image credit: Michael Drakopoulos.

Robert P Crease is a Professor in, and the former Chair of, the Department of Philosophy at Stony Brook University. He has written numerous books and articles on the philosophy and history of science, and specializes in the history of post-War scientific institutions. For 23 years he has written a monthly column for *Physics World*, called 'Critical point', on the philosophical, historical, and social dimensions of science. From 2014 to 2021 he was co-Editor in Chief of *Physics in Perspective*. In 2021 he was awarded the William Thomson, Lord Kelvin Medal and Prize (Institute of Physics) for 'describing key humanities concepts for scientists, and explaining the significance of key scientific ideas for humanists'. Crease is a Fellow of both the American Physical Society (APS) and the Institute of Physics (IOP). His last book is *The Leak: Politics, Activism, and Loss of Trust at Brookhaven National Laboratory*, with Peter Bond (2022, MIT Press).

IOP Publishing

Philosophy of Physics
A new introduction
Robert P Crease

Chapter 1

The scientific workshop

Scientific research is permeated by assumptions that are invisible to ordinary scientific theorizing and data-collecting. What assumptions about proof were at work in the discovery of parity violation or the Higgs discovery? Why did it take so long for physicists to realize that incoming neutrons caused nuclei to split rather than to absorb or shed particles? What ontological assumptions are built into most interpretations of quantum mechanics that require unreasonable conclusions about reality, and what consequences follow from rejecting such assumptions?

How philosophers implement inquiries such as these can be seen by considering scientific activity as taking place inside large organizational structures that can be thought of as a series of loosely connected 'workshops'.

1.1 Workshop frames

By workshop I don't mean necessarily an actual laboratory, only a carefully supervised and regulated environment in which special things can be prepared and observed: subatomic particles and their interactions, superfluids, chemical elements and reactions, protein folding, plant uptake of nutrients and toxins, and so forth. These phenomena do not appear in the surrounding world, or only crudely, rarely, and not in clear enough ways to study well. But in the special conditions of the workshop they can be made to show themselves for study and measurement.

While the modern world is heavily dependent on scientific workshops, they essentially originated in the 17th century. Each workshop involves 'frames', or sets of assumptions and concepts that define the physical system under study. The frames remove what appears in them from the dense network of the world outside the workshops, and prioritize what appears in them over unframed experience of that world. These frames are built on pre-judgments about the phenomena. Frames are chosen based on the phenomenon to be inquired into.

In the Newtonian workshop, for instance, frames model objects of study as masses in motion. The phenomena that show themselves as objects of study are

doi:10.1088/978-0-7503-2636-0ch1   1-1

masses, what these masses do is move about, and what starts, stops, or otherwise affects these motions are forces. The frames do not simplify objects as being *only* masses in motion, but pre-determine the class of phenomena that appear in the frame, and lay out appropriate research trajectories. The philosopher Thomas Seebohm called such framing the 'first abstraction' that constitutes the objects of modern science, one that introduces and establishes the split between primary and secondary properties, which is often thought of as the distinction between objective and subjective characteristics (Seebohm 2015, p 222).

In the Newtonian frame the table appears as a set of molecules, the body as a network of organs, and swings and chandeliers as pendulums. Such objectification is powerful and can create the impression that nature is being seen 'as it is' rather than only as it is seen in a controlled and supervised environment. What appears in workshop frames can come to seem like nature—like reality itself—and workshop activity and knowledge can come to seem like the form of all genuine activity and knowledge.

The workshop frames allow scientists to be in near-complete control of the things and events they prepare and observe, and help to understand 'the complicated array of moving things', as Feynman says at the beginning of his lectures on physics (Feynman *et al* 1963, section 2-1). The supervised and controlled conditions of the workshop can make researchers reasonably sure that the results that appear in the workshop's frame are general and do not depend on features of the world outside the workshop. Inside the workshop, researchers can put questions to nature, in Galileo's words, or question it like a court witness, in Kant's—if, that is, one conceives 'nature' as what appears in the frames. Nature is silent unless probed. The answers to these questions can then be used to help understand the surrounding world and effect changes in it.

1.2 Workshop mediations

When we enter a workshop, however, we are not stepping outside our everyday world and entering a magical space where we confront 'nature' directly. We have constructed the workshop ourselves, we have planned the projects we want to do according to our current ideas, we have built the instruments we will use with materials and designs at hand, and we interpret what we find with the concepts we already possess. Nobody asks blank or formless questions in the laboratory—scientists always pose their questions 'from somewhere', and these questions are given specific form by an inherited set of concepts, convictions, images, metaphors, habits, and equipment that make meaningful both the questions and the responses. This inheritance is not a string of biases, but what makes inquiries meaningful to scientists.

Included in this inheritance are certain fundamental and generally unquestioned assumptions about matters taken for granted in the quest for knowledge. Ambiguities, puzzles, and incongruities can arise between what appears in the workshop frames and scientists' expectations, and may call into doubt aspects of the inheritance. Workshop research proceeds by adding to the conceptual or practical inheritance—the concept of chemical element, or of black holes, or of the Higgs field, or of the molecular approach to genetics—or by seeking out and discovering some piece of evidence that the augmented tradition says should be there: new elements, black holes, the Higgs boson, the molecular structure of DNA. Science grows, that is, through an ongoing cycle of reinterpretation and inquiry.

Because of these reinterpretations, the workshop activity and its frames are historical. The frames therefore mediate the questions put, and the laboratory experimental performances are staged with a limited and finite set of 'props' and materials. What shows itself in the workshop as 'nature' is therefore antecedent to the workshop and always more than what appears there. (That what shows itself is *not* more than what appears is the basis of the interpretation of science called 'social constructivism'.)

Philosophers are interested in the ongoing interactive activity of the workshop, but look at it in a different way than do scientists, for philosophers seek to understand not what physicists know but how they engage with what they know. As the Harvard University philosopher Steven Shapin put it, 'Plants photosynthesize, plant biochemists are experts in knowing how plants photosynthesize, [while] reflective and informed students of science are experts in knowing how plant biochemists know how plants photosynthesize' (Shapin 2010, p 38). In other words, the world studied by science researchers includes not just objects but also connections between scientists and objects. This is not just an epistemological insight, for how objects of study reveal themselves in a workshop depends in part on its frames, on how its frames evolved, and on the way of life which finds it important to frame the encounter with nature this way.

Scientists have little to gain from philosophy of science when the workshop activity is clear and effective, and when conflicts between expectations and what appears in the frame is likely to be resolved by more scientific activity.

But philosophical analysis, however crude, becomes unavoidable when such conflicts cannot be resolved with more theoretical and practical scientific research. While what appeared in the Newtonian frame could be more or less reconciled with what appeared in ordinary experience—Newtonian mass, force, time, and space were not that remote from everyday encounters—things changed radically with the two 20th century epistemological and ontological bombshells, relativity and quantum mechanics. Many phenomena, including time and space, now showed up in workshop frames so radically different from the way that they do in ordinary experience as to cry out for philosophical clarification.

1.3 Workshop permeability

This made it even more urgent to address the difference between 'what scientists say' and 'what our ordinary experience tells us'.

Suppose I say, 'I am thinking', and someone replies, 'You are mistaken. Your neurons are firing.' Or I declare, 'I'm going to measure how fast this marble falls', and am told, 'You are mistaken. You're seeing matter and gravitational fields interact' (or, 'It's space-time telling matter how to move according to the curvature that matter gives it'). Or suppose I complain, 'I'll need some time to think this through', and the response is 'You can't, time is an illusion.' If the latter conclusions are indeed the conclusions of workshop science, how can it be related to my own experience in a way that does not foster either scientism or skepticism?

The disparity between the two types of questions are even present inside the workshop. Scientists in workshops experience themselves thinking, encountering moving objects, and taking time, just as they do outside workshops. 'Taking time' here means something different than what was said above about workshop activity being historical—it means the stretched-out process in which we accomplish something through projecting what we already know. I build a device by taking up my previous experiences with its parts and plans and applying them to what I want to have in the future in an unbroken, flowing process that can only arbitrarily be broken down into a series of moments. Philosophers call this process *temporalizing*. While temporalizing, and other processes to be discussed in this book, are ordinary and even essential inside the workshop, they are not analyzable with the usual scientific theorizing and data-collecting.

The sharp but easily unnoticed distinction between measured time and temporalizing is illustrated in the remark of a Paris Observatory astronomer in reaction to the recent proposal to eliminate the 'leap second', or the brief interruption in measured atomic time in order to bring it into alignment with the slightly slower rotation of the Earth. 'Now we see really closer the moment to have continuous time', said the astronomer. 'And this is something we have been dreaming about for a long, long time' (Mitchell 2022).

Yet ultimately, philosophers of science inquire not in order to assist workshop processes but to inform their own inquiries into broader issues of which the workshop and its activities are only a part.

Philosophers inquire into workshop activity mainly through three general approaches. None of these criticize, undermine, or lessen the importance and value of science, nor of the things that turn up in it—these approaches make it possible to understand how scientific thinking works. Any comprehensive account of philosophy of science must begin by describing these three approaches and showing how they are related. Each is needed for a full philosophical treatment of such a consequential human enterprise.

References

Feynman R P, Robert B L and Sands M 1963 *The Feynman Lectures on Physics* (Reading, MA: Addison-Wesley)

Mitchell A 2022 Scientists around the world vote to retire leap second *New York Times* 21 November p A8

Seebohm T M 2015 *History as a Science and the System of the Sciences: Phenomenological Investigations* (New York: Springer)

Shapin S 2010 *Never Pure: Historical Studies of Science as if It Was Produced by People with Bodies, Situated in Time, Space, Culture, and Society, and Struggling for Credibility and Authority* (Baltimore, MD: Johns Hopkins University Press)

IOP Publishing

Philosophy of Physics
A new introduction
Robert P Crease

Chapter 2

Three philosophical approaches

Philosophers have traditionally examined scientific workshops in several different ways, which can be crudely parsed into three. One analyzes the *results* of inquiry, another the *process* of inquiry, and still another *inquiring*, or what it means to *be* an inquirer.

Each approach puts a different feature of science center-stage—its logic, practice, and being-inquiring—questions it in different vocabularies for different ends, and ends up with different kinds of conclusions. The first is most closely involved with sorting through the technical details of science, the second focuses on the puzzle-solving activity, while the third connects science more with traditional philosophical concerns in exploring it as a *human* activity. Each approach sees the relevance and urgency of philosophical attention to science differently.

The approaches do not correspond to specific philosophical movements, and individual philosophers may combine aspects. The approaches differ stylistically and methodologically, although ingredients of each are blended together with history of science in what has become known as science studies, which approaches science from the start as a cultural product. Each approach has its own value, and only becomes specious with the claim that it is the *only* philosophical orientation to science. For a rough analogy, imagine the absurdity of someone saying that every problem in physics has to be approached as a quantum problem, or a thermodynamic problem, or from some other physics perspective.

My description of these approaches may sound dated, as I often quote from the founders of these philosophical orientations. But I do so because they often express most directly what motivates and continues to drive those positions. Today's thinkers may claim that they have 'surpassed', or are more nuanced or sophisticated than, the original or naïve versions of the approach with which they are most closely associated, but these approaches still operate in philosophical thinking about the sciences. This chapter will first outline them; subsequent chapters will clarify them further by showing how they address specific scientific issues.

2.1 Frame contents (OPA)

Philosophers of the first approach position themselves as onlookers, alongside scientists, of the contents of the frame. Such philosophers share the naturalistic orientation of scientists but are chiefly interested in logic and conceptual rigor, and in adopting a critical perspective on the network of products of scientific inquiry. This orientation looks at scientific findings considered apart from the inquirers or the inquiry processes that led to them and seeks the formal structure of the network of scientific truths and practices. This traditional approach is the one easiest to recognize as 'philosophy of science', and the one that is most often seen in books, journals, and conferences on the topic. It may be called the orthodox philosophical approach (OPA).

The foundation and character of this approach was set forth by the German philosopher of science Hans Reichenbach (1891–1953). Knowledge is a 'concrete thing' provided by the sciences, he wrote, and philosophers are interested in the 'internal' as opposed to 'external' relations of the contents of science. Examples of external relations are, say, how and why astronomers build telescopes and physicists accelerators, or how and why researchers evaluate the findings of these instruments. Another external relation is the 'volitional decision' leading someone to come to practice science. External relations 'do not enter into the content of science' and are to be studied by psychologists or sociologists.

The OPA prioritizes what appears in the frame over human experience, and in fact seeks to understand the latter in terms of the former. A good example is 'time'. 'We shall pay no attention to the psychological characteristics of the experience of time', Reichenbach writes. The everyday experience of time must be corrected by the physicists' notion of time. Examining the 'relativistic concept of time' will teach us, 'better than through a phenomenological analysis, what we "actually mean" by the experience of time' (Reichenbach 1958, p 113).

The goal of philosophy of science, for Reichenbach, is the 'rational reconstruction' of the contents of science, arranging them in a consistent system and finding their 'justifiable sets of operations'. It does not matter if 'actual thinking does not conform' to the reconstruction, for it arrives at a 'better way of thinking than actual thinking' (Reichenbach 1938, p 6). The aim of reconstruction is to separate the *context of justification* of scientific findings from the *context of discovery* by which these findings found their way into the frame to begin with, given the need for discoverers to take shortcuts, work with inexactitudes, and have (subjective) motivations.

Scientists may be caught up in their activity and be confused about how they are thinking; some of their 'chains of thoughts, or operations', Reichenbach wrote, 'cannot be justified' and may benefit from an outside, fully logical perspective (Reichenbach 1938, pp 3–10). For scientists absorbed in their technical activity tend to be unconscious of how they practice, and when they become aware of it cannot help 'feeling a little uneasy', for they then realize they are walking 'on the thin ice of a superficially frozen lake', and 'might slip and break through at any moment' (Reichenbach 1944, p vi).

Such philosophers start with the language of scientific theorizing and try to figure out the logical conditions for its successes. One of their principal topics concerns the reality of what appears in the frame—how and whether it is relatable to real-world things—therefore with scientific realism or idealism. They tend to share the view that concepts and theories are what can be known about the world, and that these are to be judged by testing them against observations, treating experimentation as the activity that precedes the emergence of theories and subjects them to evaluation (although some soon came to realize that observations were ordered and interpreted by theory, thus that experiment and theory worked in tandem). Just as an already-structured world is out there, assumes the OPA, so is the logic by which that structure is to be discovered. This approach focuses on the epistemology of science—on its conceptual and methodological difficulties, on how evidence is produced and evaluated, on the logic of scientific inquiry, and on the conceptual structure of its contents. Here's Reichenbach again, articulating fundamental assumptions still shared by many philosophers of science who do not strictly self-identify as of the OPA:

> [E]pistemology does not regard the processes of thinking in their actual occurrence; this task is entirely left to [sociology and] psychology. What epistemology intends is to construct thinking processes in a way in which they ought to occur if they are to be ranged in a consistent system; or to construct justifiable sets of operations which can be intercalated between the starting-point and the issue of thought-processes, replacing the real intermediate links. Epistemology thus considers a logical substitute rather than real processes (Reichenbach 1938, pp 3, 5).

While rational reconstruction of scientific knowledge is the 'descriptive task' of this approach, its 'critical task' is to use the reconstruction to evaluate the way scientists themselves construct their knowledge. Many contemporary philosophers of science who adopt this approach often acknowledge that it is impossible to fully and cleanly separate science from historical and cultural factors. Still, within this approach such factors are generally regarded as something optional, or if not optional, these factors should be something that scientists should be working to overcome and eliminate.

The spirit, practice, vocabulary, and goals of the OPA approach continue to influence mainstream philosophy of science, partly because many of its founders were logicians and scientists themselves. Philosophers adopting this approach may be said to regard physicists as the logicians and epistemologists of the world.

2.2 Frame changes (IPA)

In a second approach philosophers seek to understand how scientific researchers inquire, and how such inquiry can motivate them to radically change the frame by the objects that appear in and only thanks to it. Science is not ontology-seeking or epistemology-pursuing but *puzzle-solving*, whose prize is the payoff. This approach therefore does not attempt to separate internal and external factors, and the context of justification and the context of discovery. We may call this the instrumentalist philosophical approach (IPA).

To extend Shapin's metaphor, if students of science study how biochemists solve puzzles about how plants photosynthesize, they also find themselves having to study how biochemists have to keep devising and revising the conceptual and practical tools by which they do so. For the concepts and practices that biochemists have inherited may need to be reinterpreted, or transformed, in the light of what the biochemist puzzle-solvers encounter.

The IPA spirit is embodied in the first sentence of the philosopher Thomas Kuhn's *The Structure of Scientific Revolutions*: 'History, if viewed as a repository for more than anecdote or chronology, could produce a decisive transformation in the image of science by which we are now possessed' (Kuhn 1962, p 1). Kuhn continues, 'That image has previously been drawn, even by scientists themselves, mainly from the study of finished scientific achievements as these are recorded in the classics and, more recently, in the textbooks from which each new scientific generation learns to practice its trade.' The decisive transformation wrought by study of the history of science, Kuhn famously concluded, requires distinguishing between two kinds of scientific research. 'Normal science' consists of puzzle-solving within a frame, but particularly difficult puzzles, irresolvable within the frame, may lead to 'revolutionary science', in which findings can be made to interlock, and the puzzle solved, only by changing the frame. What then changes is not just the appearance of new things and theories, but how one theorizes and the nature of what one theorizes about. In each case the end of scientific inquiry is puzzle-solving. Philosophers of the IPA are aware that how one solves puzzles depends on the available research, and on the background, character, motive, emotions, and relations with others of the puzzle-solver.

The IPA examines the puzzle-solving process of science and human life in general, and its role in shaping human experience more broadly. Again, a good example is 'time'. John Dewey, for instance, considers the 'present' not an isolated moment to be contrasted with moments ahead and behind me, but already in process within the puzzle-solving activities I am engaged in, whether scientific, economic, artistic, political or of the most ordinary sorts:

> The situation that I am determining when I attempt to decide whether or not I mailed a certain letter is a 'present' situation. But the present situation is not located in and confined to an event here and now occurring. It is an extensive duration, covering past, present and future events. The provisional judgments that I form about what is *temporally* present (as for example in going through my pockets now) are just as much means with respect to this total present *situation* as are the propositions formed about past events as past and as are estimates about ensuing events (Dewey 1938, p 228).

In a workshop, for instance, what I am doing now carries forward past events, while the future is 'waiting for me' as the successful solution to the puzzle, or as the frustrating 'not yet' of a solution. Scientists involved with the search for the Higgs boson, or for why tau and theta particles appear to be the same and not the same, for instance, are as much in the past (or in what's driving their inquiry) and in the future (or what would bring it to a close) as in the present. Past, present, and future are not isolated moments succeeding one another but connected in and by the puzzle itself.

IPA philosophers are aware that humans do not spring into being as scientists but apprentice to become them, in the process acquiring an inheritance of concepts, convictions, and habits only in the light of which can a finding appear anomalous—a puzzle to be solved—in the first place. Such philosophers tend to share the views that inquiry involves doing rather than just cognition, that theories are descriptions of entities found in scientific practice, and that scientific work is to be judged by how well it explains, predicts, and gives us power over—rather than merely describes—the world. Philosophers of this more pragmatic tradition treat scientific activity and its results as historically based. They find that every attempt to discover timeless structures only reveals the historical character of the science that led up to it, and is revealed as a historical attempt in turn by the science to follow. Philosophers of the IPA know that what appears to be the Timeless Character of Reality is what is only Very Stable for a While.

Those who study frame changes tend to position themselves outside the workshop, while respecting the process of inquiry that takes place in it. They tend to view scientific inquiry as a puzzle-solving process that resembles the various ways of puzzle-solving in the world outside the workshop. These philosophers may be said to view scientists as puzzle-solvers of the world.

2.3 Framing (PPA)

A third orientation focuses on the *framing*, on the way of being involved in scientific inquiry; what leads people to become and remain inquirers in the first place as opposed to how they inquire or what the inquiry has found. It can also include studying the reciprocal impact of framing on human experience. I'll call this the phenomenological philosophical approach (PPA).

For scientific inquiry—puzzle-solving—is only one way to engage the world, and not the default setting; humans do not naturally or normally approach plants as biochemists, the night sky as astronomers, or sunsets as meteorologists. Humans engage the world in many other ways, valuing such things as wealth, fame, pleasure, companionship, and happiness—doing so as children, adolescents, parents, merchants, athletes, teachers, and administrators. All these modes of engaging the world arise through modifications of what philosophers call the 'lifeworld', the matrix of ways by which human beings practically connect to the world that precedes any cognitive understanding and representation.

Puzzle-solving involves adopting a certain attitude towards the world. When somebody approaches sunsets and sunrises to ask 'What's moving here?', or rainbows to ask 'What produces the colors?', or starlight to ask 'What is that it from?', or the human body to ask, 'What are its chemical constituents?', this involves taking up what we may call a *theoretical attitude* that both isolates these issues from the rest of human experience and involvements with nature, and takes up one attitude towards the world rather than others.

A *New York Times* article of 25 September 1990, for instance, quotes an engineer who was working on a device to tap energy from the waves striking the rocky western coast of Scotland: 'You watch that tremendous power all the way down the coast, and it makes you think of the hundreds of megawatts that are being dashed on the shore and not being used. The ocean is like a big battery, a huge collector and we can tap it in many places' (Simons 1990). This, of course, is the view of one person, but it's a culturally and historically fostered mode of viewing the ocean and the Scottish coastline shared by many engineers without being shared by other kinds of observers. The German philosopher Martin Heidegger devotes many pages to describing this mode, and how it arises from and is grounded in the lifeworld (1962, #15–18).

Examining the theoretical attitude does not entail that one criticize, undermine, or lessen the importance and value of it, nor of things that it turns up; the theoretical attitude, after all, is the appropriate kind of thinking for natural scientists. The philosophical aim is to understand how it arises and how its assumptions affect what shows up in it. Such an understanding is an indisputable part of a comprehensive philosophy of science.

Philosophers of the PPA therefore look at their subject differently than the way suggested by Shapin's metaphor, for they examine something other than how plant biochemists puzzle out the problems that they face; they examine—to continue the metaphor—what it is to be such a puzzler in the first place, and how biochemists interpret what they are doing in viewing plants as things that can be studied 'objectively' at all. Philosophers of science who study framing tend to turn their attention still further from technical aspects of scientific practice, focusing it more on how one comes to approach the world as a biochemist, astronomer, or meteorologist; what happens in the mode of being in which humans inquire into nature. For being a scientist is as much a particular way of being as an artist, athlete, or business person. Each has a different mode of being that shapes the horizon in which the world 'turns up', so to speak, with a different implicit understanding and valuing of the things disclosed in it. '[K]nowing is already "a" way of existing—one that is both technically powerful and socio-culturally dominant but not ontologically basic' (Scharff 2022).

Inside a scientific workshop, it is easy to forget that the theoretical attitude is only one way to disclose the world. The theoretical attitude can come to be interpreted as *the* human mode, *the* standard with which to judge human activity and experience. But the theoretical attitude involves a particular kind of disclosure, giving a primacy to things that appear in a certain (framed) way—to things that can be measured and manipulated—and tends to ignore things that do not, such as the powerful prescientific metaphors, images, and deeply embedded habits of thought in the lifeworld that shape our thinking. One consequence of this theoretical interpretation is that the scientific workshop comes to be regarded as sectioned into main halls and anterooms, with the scientific contents located in the former and things such as discussions, conferences, experimental equipment—and of course the 'volitional decision' to build and inquire in a workshop at all—in the latter. But one already has to be in the world, involved with it, living through it, in order to be able to make sense of it, and to pull out different aspects as relevant or not—let alone to inquire scientifically. As the German philosopher Edmund Husserl, the founder of phenomenology, wrote:

> The sciences build upon the lifeworld as taken for granted in that they make use of whatever in it happens to be necessary for their particular ends. But to use the lifeworld in this way is not to know it scientifically in its own manner of being. For example, Einstein uses the Michelson experiments and the corroboration of them by other researchers, with apparatus copied from Michelson's, with everything required in the way of scales of measurement, coincidences established, etc. There is no doubt that everything that enters in here—the persons, the apparatus, the room in the institute, etc—can itself become a subject of investigation in the usual sense of objective inquiry, that of the positive sciences. But Einstein could make no use whatever of a theoretical psychological–psychophysical construction of the objective being of Mr Michelson; rather, he made use of the human being who was accessible to him, as to everyone else in the prescientific world, as an object of straightforward experience, the human being whose existence, with this vitality, in these activities and creations within the common lifeworld, is always the presupposition for all of Einstein's objective-scientific lines of inquiry, projects, and accomplishments pertaining to Michelson's experiments. It is, of course, the one world of experience, common to all, that Einstein and every other researcher knows he is in as a human being, even throughout all his activity of research. [But] precisely this world and everything that happens in it, used as needed for scientific and other ends, bears, on the other hand, for every natural scientist in his thematic orientation toward its 'objective truth', the stamp 'merely subjective and relative'.

Husserl continues:

> But while the natural scientist is thus interested in the objective and is involved in his activity, the subjective-relative is on the other hand still functioning for him, not as something irrelevant that must be passed through but as that which ultimately grounds the theoretical-logical ontic validity for all objective verification, i.e., as the source of self-evidence, the source of verification. The visible measuring scales, scale-markings, etc, are used as actually existing things, not as illusions; thus that which actually exists in the lifeworld, as something valid, is a premise (Husserl 1970, pp 37–8).

The other approaches may consider things such as 'theory', 'discovery', 'explanation', 'experiment', and so on, as issues that can be detached as themes for special inquiries or book chapters. But the PPA sees them as not so detachable but part of one process, and that process as one way that humans can live. Science does not happen by adding all of these things together; science does not come in pieces like a car. The PPA considers science as a way of life that values solving the puzzles of nature in the first place, a way of life in which nature appears manipulable and measurable, and full of puzzles to be solved. Even if one could add the pieces of science together, it could be done only by having beforehand a plan about what science is.

The concept of time, once again, serves to illustrate the difference from the others. Certain scientists of the OPA proclaim that their research shows that 'Time Does Not Exist', or that it is an illusion (Rovelli 2016, ch 7). But while time may not show up as variables in their theories, the scientists who constructed such theories first had to conceive, come to understand, discuss, debate, and write out these theories. In order to be scientists, in short, these individuals engaged in temporalizing activities —a temporalizing that is intrinsic to what it is to be human, and *therefore* not restricted to puzzle-solving. What the PPA calls 'lived time' is the foundation for measured time; the latter depends on and arises out of the former.

Philosophers of science of the PPA stand outside the workshop and see scientists as caught up in the theoretical attitude in their ongoing dialogue with the world. Such philosophers are also reflective about their own mode of being, and the habits and assumptions they are bringing to bear on, and how these affect what they see. If such philosophers look at scientific looking, they look at the same time at their own looking. Philosophers of this mode may be said to view physicists—like all others and themselves—as disclosers of the world through a particular, theoretical, mode of being.

****

The above has not sought to develop specifics about particular philosophers or their work but rather to outline basic approaches. These three approaches are difficult to compare directly, as they set their focus on different dimensions of science and address them using different vocabularies. These approaches also have different views of the urgency for philosophical attention to science: to clean up its logic, to describe its puzzle-solving as a model of inquiry, to understand the connection between scientific inquiry and the lifeworld, and to head off misinterpretations of the knowledge it produces.

These orientations can be best appreciated by seeing how they approach specific topics and issues in physics.

References

Dewey J 1938 *Logic: The Theory of Inquiry* (New York: Henry Holt)

Heidegger M 1962 *Being and Time* (New York: Harper and Row) transl. J Macquarrie and E Robinson

Husserl E 1970 *The Crisis of European Sciences and Transcendental Phenomenology: An Introduction to Phenomenology* (Evanston, IL: Northwestern University Press) transl. D Carr

Kuhn T 1962 *The Structure of Scientific Revolutions* (Chicago, IL: University of Chicago Press)

Reichenbach H 1938 *Experience and Prediction: An Analysis of the Foundations and the Structure of Knowledge* (Chicago, IL: University of Chicago Press)

Reichenbach H 1944 *Philosophical Foundations of Quantum* (Berkeley, CA: University of California Press)

Reichenbach H 1958 *The Philosophy of Space and Time* (New York: Dover)

Rovelli C 2016 *Seven Brief Lectures on Physics* (New York: Riverhead)

Scharff R 2022 On making phenomenologies of science more phenomenological *Phil. Technol.* **35** 1–22

Simons M 1990 Ocean waves power a generator *New York Times* 25 September p C1

Chapter 3

Method

'The scientific method does not exist', writes the historian of science Henry M Cowles at the beginning of his book *The Scientific Method: An Evolution of Thinking from Darwin to Dewey*, 'But "the scientific method" does' (Cowles 2020, p 1).

'The scientific method', which exists in numerous versions, is a stepwise-like procedure said to characterize either how science ought to be carried out (hence, legislative), or the bare structure of how it really is carried out shorn of contingent happenings, accidental encounters, wrong terms, and subjective elements (logical). It is taught in schools, affirmed to politicians, invoked in scientific papers, and analyzed in philosophical articles.

The scientific method, by contrast—how the activity actually unfolds—lacks systematic mechanism. In private conversations, memoirs, interviews, after-dinner talks, and informal occasions scientists reveal the decisive role of accidents, memories, chance encounters, available equipment, funding, and guesswork. 'There is no such thing as the scientific method', wrote the eminent scientist and science administrator Vannevar Bush (Bush 1951). Philosopher Paul Feyerabend devoted an entire book to driving home this point, which he entitled *Against Method* (Feyerabend 1975).

It's well-known that scientists represent their work process differently to others and amongst themselves; that a gap exists between the method and 'the method'. In *The Statue Within: An Autobiography*, the French theoretical physicist François Jacob referred to it as the gap between his 'day science' and 'night science' (Jacob 1995). Historians discuss the gap under the rubric of the 'rhetoric of science' (Yeo 1986). The OPA cements the gap in the distinction between the context of discovery and the context of justification.

As each of the three orientations focuses on a different aspect of the scientific frame—on its contents, inquiry, and genesis—they each have a different conception of method. I am not concerned here with details of these approaches to method, but with showing how they relate—again, directed at an audience of physicists interested in philosophical approaches but not particularly acquainted with any one.

3.1 Frame contents (OPA)

The orthodox orientation is concerned with how a scientific method could be formulated apart from any particular application. How does scientific practice—the process by which new pieces of knowledge appear in the frame—look after application of the distinction between context of discovery and context of justification? The contents in the frame can be made to show up there one way in reality, even in different ways for individual researchers, but described as belonging there—justified—in a formalized way that is independent of all researchers. What rules can be specified beyond and above the scope of any particular scientific activity for how it is conducted—the way, say, a sports game has rules that govern each game rather than being determined by how the game goes?

Philosophers have proposed many different models for how this can be done. A simple and traditional formalization is given in *A Philosopher Looks at Science* (Kemeny 1959), a classic OPA text (dedicated to Albert Einstein, who had died just a few years earlier) by the scientist and philosopher John G Kemeny. Our 'proper subject matter', he writes—'the nature of Science'—is its method, and the 'first thing' about science to study. The 'one basic method' common to all science is to start with a question about certain facts, move to a theoretical level that can make predictions, obtain more facts through experimental inquiry, uncover further questions from these facts, and so the process continues. This is the deductive-nomological (D-N) model. Kemeny pictures it as resembling a soccer goal. Induction (turning data points into generalizations) is a post that rises from the field (or 'world of facts'), connects with a crossbar or theoretical realm—what the post rises to and terminates in—in which deductions and predictions are formulated, leading to a specific deduction or prediction—the post at the other end of the crossbar—whose verification is its connection with the ground. Science's strength depends, Antaeus-style, on regular earthly contact. As science is endless, Kemeny says, 'we may expect this cyclic process to continue indefinitely' (Kemeny 1959, pp 85–6).

Kemeny's approach assumes the natural attitude: there's a pre-existing, structured world out there independent of human thinking, we know some of it, and we learn more of and about it by careful observation, discerning guesswork, making tentative generalizations about what we know, tests, and more observations. His approach places prediction and verification center-stage, and, while it does not deny other aspects of the scientific life and practice, it assigns the study of these to history and psychology. To use another sporting analogy, the OPA views science as about scoring, and seeks an optimal scoring strategy rather than issues such as how the game evolved, the attitude of players, or its social role. The scoring strategy sought is one that can be deployed in any game, not just this or that one.

Philosophers of the OPA have developed different formalistic approaches. Besides the D-N model there is the *hypothetico-deductive* (H-D) model, which differs from the D-N model in that it does not require hypotheses to be generalizations from existing knowledge but only falsifiable; a theory becomes scientific by exposing itself to the possibility of being proved incorrect. This model, too, is not meant to be descriptive, but a philosophical test or model of what science would look like if reconstructed in logical terms.

Other descriptions of the scientific method include the *inductive-probability* model, which involves evaluations of the support that evidence provides a hypothesis, and *Bayesian* models, which assign degrees of belief to the principles that enter into a hypothesis, which in turn assigns a degree of belief to the hypothesis itself. *Bootstrapping* emphasizes the role of relevant evidence, *Evolutionary selection* recognizes the role of social factors on individuals but in a way that describes a mechanism for theory change independently of how individual researchers rationalize.

To show how Kemeny's account of OPA formalization might be applied to a concrete case, consider how the OPA would model the method of the discovery of parity violation in 1957. Two particles, the theta and the tau, had been discovered that were similar in mass and lifetime but differed in way they decayed: the theta was able to decay into two pions, the tau into three. For physicists, this was a terrible conflict—fundamental principles prevented classifying particles that behaved that way either as two separate particles or as one. Particle theory threatened to lose meaning if the particles were the same, or different. Inspired by that situation, the theorists T D Lee and C N Yang investigated, and discovered that one element of the inheritance of theoretical physics, parity conservation in the weak interaction, had yet to be experimentally verified. Learning of this, the experimenter C S Wu dropped a long-planned cruise on the *Queen Elizabeth* with her husband, on the twentieth anniversary of their exodus from China, to go to the National Bureau of Standards to stage an experiment, and her team discovered that parity was indeed violated in the weak interaction. In this case the 'prediction', modified a bit to fit the D-N model, was that experimenters would find that parity either was or was not conserved in the weak interaction (although this is already cheating, for the Lee-Yang paper only said that whether or not parity was conserved in the weak interaction was unknown and might be testable). C N Wu's work then indicated that parity was not conserved in polarized cobalt-60 nuclei. This did not by itself prove parity violation, but provoked a revision of elementary particle theory that included predictions of parity violation in other areas of elementary particle behavior, which were then tested and confirmed.

To the question of whether the actual practice of physics is described by these models, the answer is 'Of course not!' As Reichenbach insisted, rational reconstruction does not need to reflect actual thinking. That formal methods are not descriptions of how scientists practice does not necessarily matter for OPA. In many historical instances indeed, scientific activity does not easily conform to such models, particularly in observational disciplines such as astronomy and non-mathematized disciplines such as ecology. When this happens, philosophers of science often modify the above models to make them applicable to the discipline, and make some reference to how science is actually practiced (Zammito 2004). Still, such modification means that such philosophers do not, ultimately, accept the strict OPA and their work reflects more the IPA.

Consider a case involving contemporary physics, the scientific status of string theory. String theory sprang from physics theorists' attempts to avoid the problems of point-like particles by giving them area and turning them into one-dimensional strings. String theory has extraordinary scientific standing, as measured by indicators such as numbers of articles published in high-impact scientific journals, positions of practitioners at universities, talks given at conferences, prizes and awards given out, and so forth. But by other indicators, including many of those specified by the above models, string theory is an unscientific sham, its principal flaw being the fact it has no experimentally testable predictions and, barring a miracle, will have none for decades. For this reason, some scientists have (perhaps half-) seriously compared the scientific status of string theory with that of intelligent design (Ehrlich 2006). Of the theories that reach beyond the standard model, 'none of them has found empirical confirmation up to now', meaning that 'control of the theoretical evolution by empirical data, a crucial element of the natural sciences, thus appears in danger' implying 'a serious crisis of fundamental physics' (Dawid 2013, pp 1, 6). As the theory that is most vulnerable here is string theory, Dawid and other philosophers have proposed amendments to make it clear that string theory fits a philosophically acceptable version of 'scientific method'.

But such amendments involve adapting the method to actual practice rather than seeking a purely logical and formal description that would be above and apart from practice. The implications of string theory are pressuring the change; it's not 'new evidence' that is motivating the change but something in the way scientists are seeking it. Again, to adopt a sports analogy, such amendments are like changing the rules of the game because of the way the game itself is being played—that is, not because someone on the outside has decided to make it a different game—instead of viewing the rules as predetermined and optimized from the outset, which was the original goal. Such amendment-making involves admitting that the fixes may not be permanent; that accounts of method may be subject to future change.

The OPA provides a sensible way to articulate the intelligibility of scientific method when concepts, theories, and equipment are standardized. It is like observing a game at a particular instant in time, taking it as a sociological fact, and describing the logic of the rules as if these rules were pre-set. To put it another way, it is like observing a football match from high up enough that one doesn't see the ball; one can see the overall movements and from them pick out the basic rules of play, but cannot see the play itself or the experiences of the players. Another important contribution of the OPA—especially in the hands of Carnap, Gödel, Tarski and others—was, via rational reconstruction, to advance the use of logical and systematic methods in the sciences. This approach is useful for some purposes, and only becomes a wrong approach if it comes with a story about how it is the full and *only* story that describes how method *must* be.

3.2 Frame changes (IPA)

The American pragmatic philosopher of science Norwood Hanson once wrote, 'By the time a law has been fixed into an H-D system, really original physical thinking is over' (Hanson 1958, pp 70–1). Philosophers of the instrumentalist approach (IPA) seek to understand that original physical thinking, the 'how' of inquiry. What is the method by which the inquirers make the discoveries of which the OPA studies the logic? Philosophers of the IPA write about practice without assuming that it has an essence; they regard practice as self-correcting without seeking an independent model by which it happens. The point of doing science inquiry lies in its consequences; how then do scientists change their practice when they encounter something they did not expect and cannot fit into their expectations?

The IPA, therefore, sees the difference between context of discovery and context of justification as only a distinction rather than as a complete separation. Were there no distinction between the two contexts, then the moment of discovery would only be historical, and how something was uncovered would be the evidence for its truth. Were the distinction absolute, then scientific discoveries would be either completely logical—something a machine could be programmed to do—or miraculous, for non-logical thinking would be unintelligible and could not be used as justification. The IPA takes a middle path, seeing both discovery and justification as significant in different ways, different modes of normative judging, and as evolving together. What one discovers provides its own criterion for its justification. The discovery of parity violation in polarized cobalt-60 nuclei indicates what has to happen to justify parity violation as a principle.

To continue the sports analogy, the IPA conceives method as how, from their particular situations at moments in actual games, individual players might go about formulating their optimal next move. In the IPA, science is a kind of puzzle-solving. Solving a puzzle may be logically reconstructed afterwards, and guided by previous successes, but that does not address what puzzle-solving practice is going forwards. Science is not a robotic process of conjecture and refutation. It involves the ability to call into question, with discernment, inherited assumptions that are elements of our background framework, thereby opening up possibilities that could not have been foreseen at the start. What we do in laboratories is both inquiry into nature and self-inquiry. Those efforts put our guesses about nature to the test and force us to reinterpret the assumptions on which these guesses are made. Theoretical 'guesses' and experiments to test them are based on assumptions that we inherit from the entire past history of science. What shows up in the laboratory may not merely confirm or falsify the guesses, but rather call into question the background assumptions from which the guesses arise, forcing us to review and rethink the assumptions.

This is an interpretive process in which scientists must judge what is reliable and promising, and what is not. This makes some scientists of strong purpose, whose very obtuseness and reluctance to be distracted by contrary empirical evidence in this interpretative process is often a source of success. This process of inquiry may lead to changes in the contents or structure of the frame.

The IPA philosophers were strongly influenced by Darwin, whose attention to method was inspired by his worry that other researchers would not regard his mixture of observing, recording, and analyzing things like barnacles, birds, and worms as rigorous enough to count as legitimate science. Darwin's 'self-consciousness about method', writes Cowles, made him 'a careful student of scientific methodology', leading him to reflect on 'how you thought what you thought' (Cowles 2020, p 64). Darwin found the answer in what he was studying—nature itself. Just as nature has a process for choosing among alternative forms of life, so do scientists for choosing theories, involving experiment and systematic observation. 'Natural selection was nature's *method* for making species', Cowles writes, and scientists have an analogous one to make knowledge (Cowles 2020, p 65). What matters in each case is what survives.

Viewed through today's lens, Darwin's notions of method are hopelessly anecdotal, psychological, and disorganized, but they inspired philosophers such as William James, Charles Peirce, John Dewey, and others in the early 20th century to develop non-formalized accounts of how scientists acquire knowledge. Each scientist is rooted in a place and time and has inherited a particular set of concepts, tools, abilities, problems, and goals. A scientist's beliefs—meaning not mental affirmations or free-floating concepts but principles of action developed through experience—are constantly tested against the world, and when they don't fit are reshaped in an endless process. Scientists are actors in rather than spectators of nature. The scientific method is a careful characterization of this process. The 'payoff' of Darwin's species was survival; of scientists, puzzle solutions.

The American philosopher and psychologist William James did not explicitly identify anything he wrote as philosophy of science, but it can be argued that it's all he did (Gavin 1978). Humans engage with the world and build it up by pursuing their interests, in the course of which their habits are continually subjected to stress tests and revised. Scientists are different in that, while most humans dislike stress and situations where their habits clash with their environment, scientists thrive on it; their profession is to actively seek out such clashes and then find ways to overcome them. 'Let us give the name of *hypothesis* to anything that may be proposed to our belief; and just as the electricians speak of live and dead wires, let us speak of any hypothesis as either *live* or *dead*. A live hypothesis is one that appeals as a real possibility to him to whom it is proposed' (James 1897). But a hypothesis can only appear to someone as a 'live' or real possibility in the light of that person's interests, which may include thoroughness, consistency, aesthetics, purity, daring, audaciousness, ambition, and elegance—not to mention motives such as desperation, simplicity, and emulation—each of which can be productive motives in the scientific life. A scientist's life does not stop with a 'live' hypothesis that produces results; the hypothesis becomes a belief that 'appears less as a solution, then, than as a program for more work, and more particularly as an indication of the ways in which existing realities may be *changed*' (James 1907). James held to a kind of ontological agnosticism but without any lack of rigor in inquiry. 'Pragmatism does not say that anything goes', writes Stanley Fish. 'It says that anything that can be made to go goes' (Fish 1999, p 307). For the IPA, the scientific method is making-go.

There's a fantasy, James says, of an ideal puzzle solver—he's thinking of survival in evolution, but applies equally to survival of ideas—namely, 'the embodiment of the highest ideal perfection of mental development', which would be 'a creature of superb cognitive endowments, from whose piercing perceptions no fact was too minute or too remote to escape; whose all-embracing foresight no contingency could find unprepared; whose invincible flexibility of resource no array of outward onslaught could overpower; but in whom all these gifts were swayed by the single passion of love of life, of survival at any price' (James 1878).

When it came to scientific research, the philosopher Charles Peirce knew whereof he spoke. One of America's most important metrologists, he not only made precision measurements while taking part in national and international scientific surveys, but studied and improved the instruments and techniques used in them. Peirce developed instrumentation that allowed him to be the first to tie experimentally a metrological unit, the meter, with a natural standard, the wavelength of light. These experiences gave Peirce a nuanced philosophical appreciation for scientific methods, and their similarities and differences from those of ordinary life (Crease 2009). Our ordinary life is shaped by beliefs (again, for pragmatists, principles one acts on rather than ideas one affirms) mostly for granted, though we sometimes find they don't work, producing friction and discontent; similarly, scientists—and Peirce had metrologists particularly in mind—have a view of the world shaped by inherited and explored beliefs but which sometimes collide with nature, creating what Peirce called the 'irritation of doubt'. Humans have four ways of addressing such doubt and eradicating this irritation: tenacity (digging in one's heals), authority (accepting someone else's declarations), seeking the *a priori* (appealing to abstract principle), and the scientific method. The latter involves accepting that you have inherited a fallible, imperfect set of concepts and tools—ones that place you in a community of others—and have to continually re-examine and reshape them, eventually leading you to another irritation of doubt and repetition of the process, in an ever-expanding enrichment of your, and the scientific community's, encounter with nature. Peirce named the scientific method of overcoming the irritation of doubt *abduction*, a kind of robust disciplined imagination: 'It is the only logical operation which introduces any new idea' (Peirce 1935, CP 5.172), and it includes 'all the operations by which theories and conceptions are engendered' (Peirce 1935, CP 5.590). Unlike induction, abduction does not confirm or disconfirm an entire theory, but suggests explanations and connections between things.

Dewey also saw the development of ideas, scientific and otherwise, with an evolutionary eye. Human beings have needs, goals, and abilities that sometimes collide with what they find, giving rise to 'problematic situations'. Scientists are different in that they are professionally on the lookout for problematic situations, where their ideas and equipment break down, and set out to transform them, developing concepts and theories as instruments in this process. 'The scientific attitude may almost be defined as that which is capable of enjoying the doubtful; scientific method is, in one aspect, a technique for making a productive use of doubt by converting it into operations of definite inquiry' (Dewey 1929, p 182).

As the founder of the Laboratory School in Chicago, Dewey sought to develop a unified educational program for students from kindergarten through twelfth grade, and distilled his observations and findings in *How We Think*. Teachers need some simplifying principle in their educational practice, wrote Dewey in the preface, and he found it in the 'scientific' habit of thought, which was far from abstract and remote to the child mind; 'the native and unspoiled attitude of childhood, marked by ardent curiosity, fertile imagination, and love of experimental inquiry, is near, very near, to the attitude of the scientific mind' (Dewey 1910a, p 72). Dewey went on to describe this habit of thought as the following:

> Upon examination, each [problematic] instance reveals, more or less clearly, five logically distinct steps: (i) a felt difficulty; (ii) its location and definition; (iii) suggestion of possible solution; (iv) development by reasoning of the bearings of the suggestion; (v) further observation and experiment leading to its acceptance or rejection; that is, the conclusion of belief or disbelief.

Belief culminates when 'the original isolated facts have been woven into a coherent fabric'. Then:

> Sciences exemplify similar attitudes and operations, but with a higher degree of elaboration of the instruments of caution, exactness and thoroughness. This greater elaboration brings about specialization, an accurate marking off of various types of problems from one another, and a corresponding segregation and classification of the materials of experience associated with each type of problem.

Several historians of science trace the origins of the formalized notion of 'the scientific method' to these paragraphs (Rudolph 2005). While Dewey intended these steps to describe organic human thinking and how it can function in education, it was all too easy to interpret them as a special way to think, and to calcify them into an instruction manual. Authors of textbooks seized on the steps as a skeletal presentation of what science was and why it worked. Dewey's steps, Cowles writes, were quickly turned into 'a symbolism of the separation of science from everyday thinking, a talisman of scientific exceptionalism' (Cowles 2020). The result, he argues, was 'a shared method that seemed less and less natural as the twentieth century wore on' though seemingly confirmed by episodes carefully selected from history of science.

Dewey would write about the familiar elements of what we now think of as traits of the scientific method, or what he called 'systematic inference' (induction, deduction, observations, the role of experiment, and so on), and he saw himself as describing a method that applied alike to ordinary and scientific thinking, although more self-consciously in the latter. 'Scientific method is not just a method which it has been found profitable to pursue in this or that abstruse subject for purely technical reasons. It represents the only method of thinking that has proved fruitful in any subject—that is what we mean when we call it scientific. It is not a particular development of thinking for highly specialized ends; it is thinking so far as thought has become conscious of its proper ends and of the equipment indispensable for success in their pursuit' (Dewey 1910b, pp 121–7). But the last thing on Dewey's mind, Cowles concludes, was to provide a stepwise model of how scientists went about their work. He aimed rather to describe reflective thought in the most general sense to detail the way scientists used thinking as an effective guide to practical action.

While the ruminations of Darwin and the early pragmatists on method had assumed that it was a way of thinking, in the end professional anxieties and the desire to elevate and insulate scientific knowledge above other kinds of knowledge transformed it into something apart from thinking, something potentially programmable into a computer. In this way 'the scientific method'—an abstract set of rules purporting to be about scientific practice—had been turned into a logo; an identifier and an attestation. Yet it is possible, Cowles argues, to reject such a view of scientific method and to think of science 'as the flawed, fallible activity of some imperfect, evolving creatures *and* as a worthy, even noble pursuit' (Cowles 2020).

In a review of Cowles's book, the Stanford University historian Jessica Riskin argued that the 'scientific method' originated, not within science itself, but 'in the popular, professional, industrial, and commercial exploitation of its authority' (Riskin 2020). Integral to this idea, she writes, was the claim that 'science held an exclusive monopoly on truth, knowledge, and authority, a monopoly for which "the scientific method" was a guarantee'.

These suggest several motives for seeking 'the scientific method' in the first place. One reason is instructional, to spell out for students and ourselves the best way to acquire knowledge. A second possibility is insecurity, the need to reassure the public, one's colleagues, or oneself of the trustworthiness of one's knowledge. A third is to promote the cultural value of science—that, in contrast with political ideology and religion, it is *the* reliable path to acquire knowledge (Laudan 1968, pp 1–38).

Philosophers of this tradition continue to have in common an appreciation for what the formalistic OPA account assigns to the anteroom: the wrestling of scientific inquirers, as individuals and as communities, with nature—with what makes a hypothesis dead or alive, without which the formalism would be empty of content.

The IPA approach involves a pragmatic conception of knowledge as opposed to an apriorist (or modifiable apriorist) one. It is valuable for many purposes; it is good to understand and be proficient at puzzle-solving. As with the OPA, it only goes astray if it comes with a story about how it is the full and only story that describes how method *must* be.

3.3 Framing (PPA)

The phenomenological philosophical approach (PPA) sees the sciences as forms of *inquiry*, a term which refers to a particular mode of interaction between a person or community and the world. Simply put, in inquiry some vague feeling of dissatisfaction with a situation or (when more explicitly amenable to articulation) some question provokes human beings to do something that may lead to an answer—but instead of focusing on the search, as per Peirce, it focuses on the *searching* as one mode of engaging the world among others. Inquiry is not the only mode that humans have of engaging the world; however, as much as scientists may have refined and specialized the layperson's method of confronting problematic situations, the layperson seldom makes a career of making 'productive use of doubt by converting it into operations of definite inquiry'. Thinking of science as puzzle-solving therefore does not capture all of what goes into scientific activity. One might say that by the time scientific puzzle-solving has begun, what motivates it is already left behind.

The PPA views inquiry as involving more than puzzle-solving, but as a process of anticipating—of expecting or being prepared for what might be the answer—and fulfilling (or not) these anticipations. Inquiry is therefore a temporal process in which one builds on what one has experienced in the past and awaits experiences in the future that may be different from what one experiences in the present. The phenomenological orientation views method as interpreting; that is, as not just solving a puzzle, but part of a temporal, puzzle-solving process in which there is not just a payoff but a change in what and how they experience. Inquiring into method then involves examining the training, framing, adaptability, and changing envisioning of the real that takes place in inquiry.

One way of introducing and displaying the role of these dimensions in method is via the idea of 'paper tools'. Ursula Klein introduced the idea of 'paper tools' to focus 'historical analysis and reconstruction on the performative, cultural, and material aspects of representations, without disregarding the goals and other intellectual preconditions of human actors'. Klein first applied the concept to understanding the impact of the practice of writing chemical formulas known as Berzelian—a new kind of sign system—that arose in 19th century chemical practice. '[C]hemists', she writes, 'began applying chemical formulas not primarily to represent and to illustrate pre-existing knowledge, but rather as productive tools on paper or "paper tools" for creating order in the jungle of organic chemistry.' Klein argued that paper tools are 'fully comparable to physical laboratory tools or instruments and share several features with them'. These features include the fact that paper tools 'are resources whose possibilities are not exhausted by the knowledge and intentions of their inventors and by scientists' attempts to achieve present goals but rather whose application generates new goals, objects, inscriptions, and concepts linked to them' (Klein 2002, pp 2, 3). Paper tools are therefore more than instruments in the conventional, pragmatist sense, but transform the inquiry itself.

Klein's idea has been applied to other areas; for instance, Michael Gordin has applied it to the early history of periodic tables to elucidate key transformations in chemical methods (Gordin 2018). David Kaiser does something similar for physics methods in *Drawing Theories Apart: The Dispersion of Feynman Diagrams in Postwar Physics* (Kaiser 2005).

'One basic premise undergirds the argument of this book', Kaiser says: 'if we are to make sense of changes and developments in modern theoretical physics, we must attend to much more than the construction and selection of "theories".' To do so, the historian and philosopher of theoretical physics need to attend to, not logic or ways of being, but to 'how most theorists have spent most of their time' in their day-to-day professional work (Kaiser 2005, p 356). Kaiser's path is to examine the development, dissemination, and use of Feynman diagrams as a paper tool.

Feynman diagrams are line drawings that represent mathematical expressions of the behavior of subatomic particles. Richard Feynman developed them in the late

1940s and early 1950s as a device to keep track of calculations of self-energy, or the way that charged particles interact with their own fields. These calculations are done by what is known as perturbation expansion, which works by viewing each self-energy interaction as a small change, or perturbation, of some known state. A perturbation calculation then tries to add up an infinite number of such small changes. Keeping track of them all, let alone adding them up, makes such calculations forbidding. In work for which he would win the Nobel Prize in Physics, Feynman used the diagrams as a 'bookkeeping device for wading through complicated calculations' (Kaiser 2005, p 43). Kaiser's account of how physicists used this imaging tool reveals several dimensions of particular interest in the PPA approach.

Training. As simple line drawings, Feynman diagrams can readily be grasped at a glance as a whole. But, as Kaiser notes, the visual simplicity is deceptive. Even physicists at first did not regard the diagrams as natural or intuitive. 'Research tools such as Feynman diagrams never apply themselves; physicists have to be trained to use them, and to interpret and evaluate the results in certain ways' (Kaiser 2005, p 6). Like other tools of theoretical physics, the diagrams 'hardly ever seem natural or obvious on their own'. Using them is 'closer in kind to craft skill or artisanal knowledge than to explicit textual information' (Kaiser 2005, p 17). It was difficult, though not impossible, for the diagrams to spread via texts alone. Kaiser compares the spread of Feynman diagrams to the spread of artistic styles, using Ernst Gombrich's famous discussion in *Art and Illusion* about the apprenticeships required in the acquisition of the techniques associated with 'realism' in painting. Painting and other art forms are practical, bodily skills where the need for apprenticeship is obvious. But Kaiser shows that apprenticeship is required even in the transmission of theoretical tools. In *The Structure of Scientific Revolutions* Thomas Kuhn emphasized the importance of students' mastering the analysis of examples and modeling in order to pass on paradigms. Kaiser finds this true in the case of Feynman diagrams as well: 'theoretical physicists do not enter the field on the basis of correspondence courses, sending and receiving written instructions in the absence of any face-to-face training interactions' (Kaiser 2005, p 11). He traces the spread of Feynman diagrams from those whom Feynman tutored in their use, to a core group who learned the technique at the Institute for Advanced Study, to those who collaborated with members of the Institute (Kaiser 2005, p 108). 'The diagram's rapid dispersion depended much more strongly on personal contact and particular pedagogical relationships than on published tests' (Kaiser 2005, p 27). The basic principle was stated in Robert Oppenheimer's quip that 'the best way to send information is to wrap it up in a person' (Kaiser 2005, p 109). Moreover, the training depended on local conditions—what problems those in that part of the workshop were addressing and how they were addressing them. 'Feynman diagrams spread most often by causal, local interactions—*this* person in the diagrammatic network mentoring *that* student in the new techniques' (Kaiser 2005, p 357). Nor is Kaiser's empirically based description of the dispersion of this theoretical tool limited to diagrammatic ones.

Framing. Mastery of the craft knowledge then produces what amounts to a frame that shapes the subject matter of theoretical physics; the PPA, that is, examines the what and why of framing. The aim of training in theoretical physics, that is, came to include the ability for the novitiate to use the diagrams to 'see' what everyone else did. It therefore affected the paper tool users, shaping the projects that they undertook and how they undertook them, as well as what projects they could imagine doing. But the frame was not arbitrary, and succeeded because it was an effective tool for carrying out the kinds of calculations for it was used. Here again there are affinities to realism in painting, where realism came to be defined by the techniques, and the techniques by the tradition of realism. The success of the diagrams also reinforced the sense that they were part of the frame; that a researcher's discovery with the aid of a tool reinforces the researcher's confidence that the tool is the correct one (in what I'll refer to later as the 'Treiman effect'). Researchers, that is, often grope about with a complex of assumptions, intuitions, and speculations that form a kind of mental map. Any discovery that a researcher makes brings new trust in this map, even when elements of the map clearly need to be adapted or changed. Research is like filling in a crossword puzzle: finding that a certain word fits solidly and securely confirms not only that word but also the validity (or invalidity) of the other penciled-in words that had inspired those words in the first place. It is easy to oversimplify the discovery process—to portray it as moving progressively from a solid base of what we know, guesses about what we don't, tests of those guesses and new knowledge added to the original base. Growing confidence in the calculations enabled by Feynman diagrams thus reinforced confidence in the diagrams themselves as a tool, thereby in the diagrams' applications, and so on.

Adaptability. In traditional history of science, or at a superficial glance, theoretical tools are thought to spread like 'batons in a relay race—stable objects that retained their meaning and form as they were passed from one user to another in a growing network'. Kaiser's close examination reveals a different story, that Feynman diagrams 'were actively appropriated and deployed in a fast-expanding assortment of ways', ways that a phenomenologist would recognize as revealing the tool's adaptability. When he first conceived them, Feynman regarded his diagrams merely as a bookkeeping tool and nothing more: 'During the 1950s and 1960s, Feynman diagrams did not compel, by themselves, a unique meaning or interpretation' (Kaiser 2005, p 5). The diagrams had to be interpreted by their users, 'and did not impose any inner meaning upon the world by themselves' (Kaiser 2005, p 173). Tweaked and refashioned, they found were applied to new applications, with physicists 'tweaking their pictorial form and reworking their relations with other calculating tools to better serve and array of tasks—many of which had not been recognized as problems, let alone solutions, when Feynman and Dyson introduced the diagrams' (Kaiser 2005, pp 173–4). None of these interpretations were predetermined, and were 'usually influenced by local experimental demands' (Kaiser 2005, p 358). First used in high-energy physics, Feynman diagrams soon found uses in nuclear physics, solid state physics, gravitational physics, and an ever-widening circle of applications. 'Improvisation and bricolage', Kaiser writes, 'can lead to applications that had never been envisioned by the tool's inventors' (Kaiser 2005, p 356).

Envisioning the real. In an old story familiar to philosophers of science and technology, tools such as Feynman diagrams, once mastered, not only shape the practices of those inside the workshop but come to be taken for granted and become seemingly transparent avenues to what appears to be the 'real'. As Kaiser writes, 'Some physicists interpreted Feynman diagrams as capturing reality more directly or completely than other visual tools did, regardless of how standard, familiar, or easy to use than other tools were' (Kaiser 2005, p 362). Such an interpretation was strongly reinforced by how Feynman diagrams were represented in textbooks. The diagrams, in fact, did not look so much different from photographs of the tracks of particles in cloud and bubble chambers, which staged much of high-energy physics in the 1940s and 1950s—the time of the diagrams birth. Such photographs were not, of course, of the particles themselves but of their traces in condensations and bubbles, but these represented the tracks of the particles themselves. These photographs and the diagrams had a 'shared visual schemata' (Kaiser 2005, p 372). One textbook even sandwiched such photographs and Feynman diagrams, 'tacitly placing Feynman diagrams within a continuous series of visual evidence about real particles' behavior' (Kaiser 2005, p 372). Sometimes the diagrams were reproduced as appearing in the lens of a magnifying class, as if the observer were actually peering into a visual phenomenon. Kaiser, again, compares this to Gombrich's analysis of traditions of realism that (falsely) appeared to represent nature to the 'innocent eye' (Kaiser 2005, p 361).

Kaiser writes, 'Feynman diagrams helped to transform the way physicists saw the world and their place within it' (Kaiser 2005, p 4). But how to understand that transformation? Feynman diagrams confirm phenomenological intuitions that the mediating influence of instruments—this time in the form of 'paper tools'—occur throughout science, and that science is impossible without it. The story of the diagrams also illustrates what becomes uniquely visible in the purview of the phenomenological approach, and in several senses. One is that empirically analyzable mediations of experience occur even in the methods of theoretical physics, mediations that do not become thematized in other approaches to science studies. Against the OPA, the phenomenological approach shows that theory is ultimately less important than the practices of mediation. Try as we might, Kaiser writes

> We will never come across a 'theory' in the flotsam and jetsam of our sources—and thus we should be wary of letting the categories of 'theory construction and selection' direct our historical analysis. Instead, when we inspect the materials with which theoretical physicists have worked, night and day, we see tinkering and appropriation of paper tools—tools fashioned, calculations made, approximations clarified, results compared with data, interpretations advanced, analogies extended to other types of calculations or phenomena, and so on. 'Theories' do not appear, nor is it clear where they might even be found (Kaiser 2005, p 377).

The PPA thus pays less attention to theory in the methods of science than more traditional philosophies of science, and pays more attention to transformations in the experience of those who inquire.

****

An amusing example of what is not the scientific method in any of these approaches took place in September 1946 in New York City at one of the first post-war annual meetings of the American Physical Society. At one session, the presentation by the young theorist Abraham Pais was interrupted by Felix Ehrenhaft, an elderly Viennese physicist. Ever since 1910, Ehrenhaft had been claiming to have evidence for the existence of 'subelectrons' whose charges were smaller than that of the electron. Now approaching 70, he was still seeking an audience, and was approaching the podium demanding to be heard.

A 24-year-old physicist named Herbert Goldstein was sitting next to his mentor and former colleague from MIT, Arnold Siegert. 'Pais's theory is far crazier than Ehrenhaft's', Goldstein asked Siegert. 'Why do we call Pais a physicist and Ehrenhaft a nut?' Siegert thought a moment. 'Because', he said firmly, 'Ehrenhaft *believes* his theory' (Crease 2001).

Ehrenhaft's proposal may have looked scientific from the outside. But it lacked a formal procedure—a formal process as per the OPA. It was also not the puzzling out of something, as per the IPA, but a claim that he wanted others to agree to. Finally, as a philosopher of the PPA would note, Ehrenhaft was motivated by conviction; by an attitude that had interfered with the normally playful attitude that scientists require, an ability to inquire—risk and respond—and to transform the inquiry. (Conviction, Nietzsche said, is a greater enemy of truth than lies.) What turns someone into a crackpot can be not a misuse of formal procedure, or failing to puzzle out something, but the disruptive effects of the author's conviction.

References

Bush V 1951 What every layman should know *Saturday Review* 17 February pp 14–5
Cowles H M 2020 *The Scientific Method: An Evolution of Thinking from Darwin to Dewey* (Cambridge, MA: Harvard University Press)
Crease R 2001 Crackpots and their convictions *Physics World* May
Crease R P 2009 Charles Sanders Peirce and the first absolute measurement standard *Phys. Today* **62** 39–44
Dawid R 2013 *String Theory and the Scientific Method* (Cambridge: Cambridge University Press)
Dewey J 1910a *How We Think* (Boston, MA: Heath and Company) preface
Dewey J 1910b Science as subject-matter and as method *Science* **31** 121–7
Dewey J 1929 *The Quest for Certainty* (Carbondale, IL: Southern Illinois University Press)
Erlich R 2006 What makes a theory testable, or is intelligent design less scientific than string theory? *Phys. Persp.* **8** 83–9
Feyerabend P 1975 *Against Method: Outline of an Anarchistic Theory of Knowledge* (New York: Verso)
Fish S 1999 *Truth and Toilets, in The Trouble with Principle* (Cambridge, MA: Harvard University Press)

Gavin W J 1978 William James' philosophy of science *New Scholasticism* **52** 413–20

Gordin M 2018 *A Well-Ordered Thing: Dimitrii Mendeleev and the Shadow of the Periodic Table* (Princeton, NJ: Princeton University Press)

Hanson N 1958 *Patterns of Discovery an Inquiry into the Conceptual Foundations of Science* (Cambridge: Cambridge University Press)

Jacob F 1995 *The Statue Within: An Autobiography* (New York: Cold Spring Harbor Laboratory Press)

James W 1878 Remarks on Spencer's definition of mind as correspondence *J. Specul. Philos.* **12** 1–18

James W 1897 *The Will to Believe* (New York: Longmans, Green)

James W 1907 *Pragmatism* (New York: Longmans, Green)

Kaiser D 2005 *Drawing Theories Apart: The Dispersion of Feynman Diagrams in Postwar Physics* (Chicago, IL: University of Chicago Press)

Kemeny J G 1959 *A Philosopher Looks at Science* (Princeton, NJ: Van Nostrand)

Klein U 2002 *Experiments, Models, Paper Tools: Cultures of Organic Chemistry in the Nineteenth Century* (Stanford, CA: Stanford University Press)

Laudan L 1968 Theories of scientific method from Plato to Mach *Hist. Sci.* **7** 1–38

Peirce C S 1935 *Collected Papers of Charles Sanders Peirce* vol 5 (Cambridge, MA: Harvard University Press)

Riskin J 2020 Just use your thinking pump! What is the scientific method, and when, where, and how did it become, as the kids say, a thing? *New York Review of Books* 2 July

Rudolph J L 2005 Epistemology for the masses: the origins of 'The Scientific Method' in American schools *Hist. Educ. Q.* **45** 341–76

Yeo R R 1986 Scientific method and the rhetoric of science in Britain, 1830–1917 *The Politics and Rhetoric of Scientific Method: Historical Studies* ed J A Schuster and R R Yeo (Dordrecht: Reidel)

Zammito J 2004 *A Nice Derangement of Epistemes: Post-positivism in the Study of Science from Quine to Latour* (Chicago, IL: University of Chicago Press) pp 6–14

IOP Publishing

Philosophy of Physics

A new introduction

Robert P Crease

Chapter 4

Discovery

'Discovery was discovered' argues David Wootton in *The Invention of Science* (Wootton 2016). European languages lacked the word prior to the 15th century, when it was 'not an established concept'. It was widely assumed that there was no need—all knowledge humans could gain about the world already existed or had been lost, forgotten, or mislaid. In the heavens nothing ever changed, on Earth there was 'nothing new under the Sun'. History was not conceived as a linear timeline on which fundamentally new things could appear to shake things up. Thus Columbus announced this finding of the New World using the word *inventio* (find out), and Galileo reported his finding of the satellites of Jupiter with the word *exploro* (explore). Both, as well as other sometimes awkward locutions, are now routinely but anachronistically translated as 'discover'.

'Discovery is not in itself a scientific idea but rather an idea that is foundational for science: we might call it a metascientific idea' (Wootton 2016, p 103). The impact of the word 'discover' stemmed from its being an 'actor's concept'. Once it's publicly available, Wootton writes, 'you can set out to make discoveries, knowing that is what you are doing', in a ratification of the theoretical attitude. The concept makes possible the scientific workshop and its activity. 'It is easy to say that our world has been made by science or by technology, but scientific and technological progress depend on a pre-existing assumption, the assumption that there are discoveries to be made' (Wootton 2016). While the origin of discovery is historical-linguistic and its practice scientific, its structure is described philosophically.

We can name this structure *recognition*, or the transition from unknowing to knowing. Recognition is usually thought of as pertaining to people; from not being able to make out who someone is to knowing who they are. But philosophers speak of the recognition of social groups, of objects and patterns, and of styles and situations. In a scientific context, recognition can be of already known natural features of the world, but also of features apprehended for the first time. For science, indeed, the distinctions are critically important between recognizing, say, an

doi:10.1088/978-0-7503-2636-0ch4

already-known galaxy, a known but forgotten galaxy, a previously unknown galaxy, and the first-time recognition that there is a feature of the heavens called a galaxy that is something different from nebulae and comets.

Again, while contemporary philosophers of science generally combine elements of more than one approach to scientific recognition, the three I have described are the basic ones.

4.1 Frame contents (OPA)

In *A Philosopher Looks at Science* Kemeny told a canned and condensed story of the Neptune discovery (Kemeny 1959). The example served beautifully: deviations between measured and predicted positions of the planet Uranus were noted; a pattern detected in these deviations; a theoretical model based on Newtonian mechanics was proposed to account for the deviations; a key element of the model was a hitherto undiscovered planet in a specific orbit; a telescope was pointed at a particular place in the sky; a planet appeared. The recognition followed a determination of the solution to the puzzle, and the location of it as predicted. Predict, look, find.

That way of telling the story intentionally leaves out the context of discovery. This was messy. At least three people, including Galileo, had observed and recorded the planet, but had assumed that it was a star (we'll call these pre-discoveries). Two centuries later, a French astronomer tracking the orbit of the planet Uranus noticed that it did not match predictions. An English astronomer learned of this and calculated the position, orbit, and mass of a heavenly body that might be causing it and alerted the Royal Observatory, which dawdled. Meanwhile, another French astronomer did a parallel calculation and publicized the news. An astronomer at the Royal Observatory noticed the similarity of the two calculations, finally took the proposal seriously, and—assuming that a French observatory was already looking for it and would scoop his own—frantically began searching for the planet in secret, but in the wrong place. The French astronomer could not interest his own country's observatory to look for the planet and, frustrated, sent his calculations to Berlin. The astronomers there *were* interested, put down their other projects, and used their telescope to observe and identify a planet. A heated controversy ensued over who discovered the planet, leaving many scientists in several countries with a range of passions including puzzlement, regret, and anger. The OPA model side-lines those personal, psychological, social, and historical dimensions of the discovery, as well as the instrumental context: the telescopes, the observatories, the techniques, the measuring and recording equipment.

Allan Franklin adopts a similar, context-of-justification oriented approach to the discovery of parity nonconservation: parity is 'suggested' not to be conserved, experiments are performed 'with positive results,' and the result was that a fundamental principle of theoretical physics was overthrown. Franklin then directs his attention to the way the evidence decided the issue between two classes of theories (Franklin 1986, ch 1).

The OPA does not concern itself with how the scientists came up with their ideas, focusing not on how something new appears in the frame but on what happens after it appears. It assumes that something really exists and is accurately described. Something that was hidden in the woods is now in the clear. The path into the clear is not a concern, although there are differing opinions on why. Hans Reichenbach, Carl G Hempel, and other early positivists took it for granted that the context of discovery was a matter for psychologists and historians, for as it depends on an unanalyzable creative intuition that cannot be turned into algorithms or formalisms. Paul Feyerabend saw no structure at all to that path, vehemently expressing his opinions in *Against Method* in what has been called the 'discovery machine objection' (Curd 1980, p 207). Other philosophers in the OPA tradition have developed what they call a 'logic of discovery', but what they mean is the set of formal tests that evaluate the reality of a new phenomenon or idea—its legitimacy in the frame. Or they suggest that discovery springs, not from algorithms or logic, but from 'scientific judgment' (Wartofsky 1978, p 9).

Philosophers of the OPA tradition, in short, do not analyze the *context* of discoveries, or how they were led out of the woods, but the *justification* of these discoveries, or how they are known to be in the clear. It relegates discoverers, their tools and methods, to the anteroom of science. As the philosopher of science Norwood Hanson put it, such a logic of discovery is more like a logic of the 'finished research report' (Hanson 1958).

Nor does the OPA consider the lifeworld, the matrix of attitudes and assumptions that first have to be in place for the discovery of something like Neptune to happen: that one knows what a planet appears like, that one cares that it's not in the spot that it's supposed to be, that one knows how to find it on the basis of the theory, that one has the instruments to do so, that one knows how to use those instruments, and so forth. The discovery process according to the OPA involves looking and finding something recognizable that appeared more or less as expected at the anticipated spot with the correct instruments.

The OPA approach does much to clarify the discovery process when the concepts, theories, and equipment are standardized, especially when complex logical methods are required. Articles in a special issue of the journal *Synthese* entitled 'A philosophical look at the discovery [of] the Higgs boson' analyze the structure of how the 2012 finding took place: the processes of data selection, how particle events were reconstructed, the techniques used in evaluating events of interest, and how all these justified the conclusion that the Higgs particle had been recognized. Other articles explored the role of theory, calibration, inference, and probability. These analyses contribute to philosophy of science, wrote editor Richard Dawid, 'by raising substantial philosophical questions with respect to the relation between theory and experiment, the epistemic status of theoretical statements and the notions of empirical confirmation and discovery' (Dawid 2017).

Discovery can indeed be portrayed artificially and for specific purposes as the outcome of applying logic. But it does not have to be, and is certainly not *only* what's involved in discovery. The acquisition of knowledge goes hand in hand with acquisition of an awareness of the complexity of the knowing process.

4.2 Frame changes (IPA)

It is easy to oversimplify the discovery process—to portray it as based on a solid base of what we know, guesses about what we don't, tests of those guesses, and new knowledge added to the original base. There's another process at work, the focus of the IPA, which focuses on the tentative and often hesitant process of inquiry in which each step leads to the next, and in which each new result affects our assessment of the path we have been taking.

The tentativeness of such inquiry is illustrated by a lesson taught by Princeton University theoretical physicist Sam Treiman. Cool and clearheaded—reserved without being remote, dedicated without being distant—Treiman knew that a researcher usually gropes about with a complex of assumptions, intuitions, and speculations that form a kind of hazy mental map. Any discovery you then make is also significant in that it brings new clarity and trust in this map, even when you find elements that must be changed.

Treiman was careful in his classes not to present scientific discovery in the usual formal and tidied-up way. Whenever he finished discussing an extraordinary discovery, he would often remark, 'So…we are invited to jump to two conclusions at the same time.' One would be the veracity of the discovery itself, the other a renewed confidence in the path one took to it. The Higgs boson discovery and the Bohr atom, for instance, did not simply give physicists another particle or an atomic structure but reinforced physicists's confidence that they had been on the right track in the first place. The *Treiman effect*, as it can be called, reflects how each new result affects researchers' assessment of the path they have been taking all along, and how the path is justified by its successes (Crease 2013). It reveals an understanding of science informed by a robust practical sense of the scientific inquiry that motivates OPA but is not a part of its ambitions. It reflects a practical sense of how science advances, of its genuinely creative character inasmuch as it opens up new paths for itself rather than insisting that the old ones are the only way forward.

In this practical sense, the discovery comes first, and only it ratifies the path taken to it as well as its justification. The IPA approach does not seek formalized rules for processes involving fulfillment of expectations, but investigates the way scientific inquiry leads to discoveries in cases where fulfillments and expectations evolve together. The approach is as interested in the path to recognition as the structure of what is recognized.

The IPA is not simply turning the OPA on its head by paying attention to the path rather than the structure. Paying attention only to the path results in a picture such as that, found in a book on the discovery of insulin, according to which it is 'a tale of bold, lone geniuses or saints who set to work on improving the lot of humanity. Instead, it is a story of monstrous egos, toxic insecurities, and bitter career rivalry that at times resembles "Game of Thrones" but enacted with lab coats and pipettes, rather than chain mail and poisoned daggers' (Hall 2022). For the very historicity of recognition helps build the structure of what it is that is to be recognized.

A classic example is the discovery of oxygen. Unlike the Neptune episode, the assumptions and concepts differed at the beginning and end; the concepts evolved with what the scientists thought they were looking at, as well as their justification. In order for oxygen to be labeled an element, the concept of element first had to emerge and justify itself. The discovery of oxygen also did not involve testing of a hypothesis, but rather repeated revisions of the concepts being applied.

'The sentence "Oxygen was discovered" misleads', wrote Thomas Kuhn. 'At least three scientists—Carl Scheele, Joseph Priestley, and Antoine Lavoisier—have a legitimate claim to this discovery', along with Pierre Bayen (Kuhn 1962, ch 6). Scheele's research was completed first but not publicized until after the work of the other two, so it played no role in the discovery. Bayen discovered that heating a mercury compound produced a gas that he described as 'fixed air', Priestley's research found that the gas could support combustion, and called it 'nitrous air', Priestley told Lavoisier about this, who redid the work and classified the resulting gas as 'common air'. Each of the three chemists therefore had identified the gas produced by heating the mercury compound with a different, already known substance. Priestley again experimented with the mercury compound, now concluding that the gas was not fixed air but 'purified' air, that is, air from which all the phlogiston had been removed. Lavoisier then re-examined that work and determined that the gas was not air but a separate component of it. Nowadays, the fact that Lavoisier recognized that a substance today called 'oxygen' is an independent constituent of air 'counts' as the discovery.

The reason the question 'When was oxygen discovered?' misleads, Kuhn points out, is that it suggests that 'discovering something is a single simple act unequivocally attributable, if only we knew enough, to an individual and an instant in time' (Kuhn 1962, p 55). If we take the question to be asking about the time someone first possessed an impure sample of oxygen, the discovery was made by the person who first bottled air. Bayen, Priestley, as well as others before them, had produced purer samples of oxygen than air, and knew that it was not normal air, but none of them thought that they had made a discovery, for each thought that they had simply discovered a new method to produce something already known. Some fastidious philosopher, Kuhn suggests, might protest that even Lavoisier did not 'discover' oxygen, for while he identified the gas as a separable component of air he misunderstood aspects of it from the modern perspective; the concepts by which it is understood have continued to evolve.

The discovery of oxygen, in the sense of the path to its recognition, was an interpretive process. As such, it exhibited six key features: *historical embeddedness, reinterpretation, temporality, invariance, showcase profiles*, and *unanticipated consequences.*

Historical embeddedness. Scientific inquiry is always 'from somewhere', given specific form by the particular assumptions, concepts, and practices that comprise the training when scientists enter the workshop. This inheritance makes meaningful both the inquiries practitioners make and what they find. Inquiry is therefore always from a particular perspective, and directed at a laboratory performance staged with a limited set of 'props' and materials, and it is impossible to jump out or get behind our historical embeddedness to evaluate how to proceed, or to check our discoveries. No search for the Higgs boson took place, or could have taken place, at any laboratory in the world in the 1950s.

Reinterpretation. Scientific practice involves constant reinterpretation of its inheritance. Interpreting the meaning of a book or work of art requires progressively revising one's understanding about it and making that understanding richer and deeper with further reading or reflection. Interpreting an experimental result—except in special, crystal-clear cases such as the initial recognition of Neptune—generally forces researchers to repeatedly rethink and revise aspects of their inherited knowledge. Mismatches between what is expected and what found may call into doubt features of the inheritance and force inquirers to revise those features. The revisions are made along with the discovery itself; thanks to such revisions, scientists *acquire the capacity* to recognize. This is why benchmarks for completion of the discovery can be unclear. To recognize oxygen, for instance, the scientists first had to arrive at the notion of an element, a separable substance from air, before recognizing oxygen as an instance. But the substance soon to be identified as oxygen helped lead to the concept of element, of which it was an example. Contemporary discoveries—such as, say, of the Higgs boson—required a century of evolving concepts and ideas, and how these are integrated. Furthermore, researchers can never be certain in advance where insights may come from that help revise assumptions and expectations. Who knew that the path to the Higgs boson required ideas about radio transmission through atmospheric plasma, or ideas from solid state and geophysical concepts? Scientists of the IPA but not the OPA are interested in the role of such revisions.

Temporality. Recognition is temporally protracted. Recognizing a person may seem to be instantaneous, as when you recognize someone coming towards you from a distance, but it is made possible by already having come to know that person as well as a growing perceptual experience. It may look like some discoveries—so-called 'Aha! moments'—are instantaneous, and in OPA accounts can be treated that way. But discoveries of that kind only happen when the discoverers know from the start what to look for, with anticipations already in place waiting to be fulfilled. When Mendeleyev first came up with a chart today known today as the periodic table, it had holes beneath boron, aluminum, manganese, and silicon. Mendeleyev decided that these holes weren't the result of inadequacies in his classification scheme, but would be filled by hitherto undiscovered elements. He was so confident that his work showed that these elements existed that he described their chemical behavior and gave them names by attaching the prefix 'eka' (from the Sanskrit for 'proto') to those they most closely resembled; the element below aluminum, for example, became eka-aluminum. When chemists looked for these elements, therefore, they were seeking and discovering something that they expected to be there. But the development of the periodic table itself, which gave them that confidence, was a lengthy process, and required the building of paper tools, which did not simply classify elements but changed how chemists conceived them, giving them new perspectives, concepts, and goals. The process needed for chemists to recognize elements was already in the past; chemists had what they needed to discover. Even the experience of something as 'missing' requires temporalizing. 'Aha! moments' and temporal interpretive discoveries are therefore not at two ends of a spectrum of recognition; the former is a special case of the latter.

Invariance. Recognition of something requires apprehension that unfolds not just over time; it also involves grasping it as able to appear as 'the same' in different conditions. To recognize a person walking towards you is to grasp that person as the same person who appears when they come closer, turn around, bend over, or walk into sunlight or dark. More generally, to experience something as existing, philosophers say, is to apprehend it as having different 'profiles' that can be experienced in different conditions. Parity violation, for instance, was such an extraordinary discovery that, when the news circulated throughout the physics community, it made physicists realize that it should have profiles in many other areas than where it was originally spotted in a particular nuclear physics experiment. In many laboratories, physicists dropped whatever they were doing to look for it, such as in bubble chamber photographs or pion decays. This is not 'replicating' the experiment, any more than seeing a cup from two different angles is 'repeating' the cup, but seeing different profiles of the same phenomenon. 'If it appears *this* way over *here*', experimenters thought, 'it must look *that* way over *there*.' This is what I mean by the justification of the discovery emerging with the discovery itself. Experimenters routinely vary profiles in calibrations and by changing things like target materials, energy, and other systematics.

Showcase profiles. Scientific objects, like other kinds of objects, are at first not seen with Cartesian clarity. One may have to adjust how we perceive them to be able to see them in a way that suits our interests more. When we see a house from a particular angle, Husserl remarked, it may inspire—and even push and compel us—to view it from another direction, so that we might have a clearer, more instructive, and more satisfying glimpse of it given our interests.

Unanticipated consequences. The American sociologist Robert Merton once devoted an essay to 'the unanticipated consequences of purposive social action', or the way actions can have consequences unintended by the actors. While Merton was referring to social acts, the point also applies to scientific discoveries. This scarcely needs repeating. Unanticipated consequences of discoveries can run the gauntlet from changing scientific research toolboxes and trajectories to major social impacts. The recognition of oxygen helped solidify the concept of element, and made possible other lines of research. The recognition of Neptune helped build the confidence in using mathematical models based on deviations from the expected positions of heavenly bodies.

A few philosophers of the IPA have described the discovery process more in detail. William James wrote about the role of factors such as curiosity, wonder, equality, and aesthetics in motivating and guiding puzzle-solving and shaping its proposed solutions. John Dewey emphasized the tentativeness of inquiry, and how its success is transformative. 'Every genuine discovery creates some such transformation of both the meanings and the existences of nature' (Dewey 1925, p 157). Charles Saunders Peirce (Peirce 1935), I mentioned, named the path to discovery *abduction*. Non-scientific judgments entering the discovery process have also been noted in the Higgs discovery (Staley and Elliott 2017).

Some philosophers concern themselves with the arc of the discovery path. While Kuhn famously drew a sharp distinction between 'normal science', which turns up results that fit snugly with the rest, and 'revolutionary science', in which the findings are sufficiently puzzling (anomalous) that they are unable to fit the frame and which force us to change that frame, or 'paradigm', the philosopher and historian of science Mara Beller sees not conceptual leaps in such a process but rather 'openness, selective borrowing, and communication' in a process of the 'intricate flux' of dialogue, whose outcome may be a collective conceptual shift. Beller found, furthermore, that such a 'dialogical and communicative' way of thinking illustrates the lesson that 'an ongoing responsiveness to the concerns of others, in addition to being a basic human value, is a precondition of scientific creativity' (Beller 1999).

The Treiman effect embodies all these features of the recognition process. To be an inquirer, in the eyes of the IPA, is to knowingly engage in a temporal and interpretive activity seeking to grasp a worldly invariant, sometimes in a showcase way, whose recognition will add to and therefore potentially transform scientific research and perhaps add to the world.

In this interpretive process a mismatch is experienced between what you are experiencing and the categories you are using to understand it, leading you to revise the categories, experience something new, further revisions, and so on. An interpretive process involves a continuous reinterpretation in response to changing experience until recognition is achieved. Such a recognition presupposes the environment or horizon in which that thing can appear, a horizon that has itself evolved. It takes a world to recognize, and it takes a world to make a discovery.

The parity violation discovery, too, embodies each of these features. It grew out of the state of theoretical physics of the time, much of which concerned discovering and classifying new particles. These classification systems all relied heavily on symmetry, and classifying the tau and theta required reinterpreting the role of symmetry. The tau–theta puzzle evolved over time, growing into a serious problem that attracted the attention of forefront theorists. Its solution was found to be observable in different contexts and physicists sought the most dramatic of these contexts. The outcome reshaped theoretical physics.

The IPA approach to discovery does much to clarify the puzzle-solving process involved in discovery, and puzzle-solvers, I think, benefit from seeing such a clarification. The Treiman effect, for instance, helps to show how misguided it is to get stuck on a formalistic approach to discovery or its justification—there are many paths to recognition. The IPA account supplements the OPA in that it is sensitive to the 'stage-setting' required for discovery that the OPA passes by. The OPA approaches discoveries as in effect dropping out of the blue, fully formed and matured, complete with justification. But the concepts involved in the discovery, as well as the means of justification, arise from and in the process itself.

4.3 Framing (PPA)

The PPA seeks neither formalized rules nor examines the process of scientific inquiry, but looks at the experience of an inquirer. Is having an anomaly enough to spark a revolution, for instance? Only for those who experience it being a problem—and an urgent one—in the first place; only for those compelled to puzzle over and pursue it. Just as the IPA sees puzzle-solving in the widest context, the PPA sees the compulsion to be a puzzle-solver in the widest context. What is it that an inquirer lives through?

According to a view long enshrined in textbooks and ratified by traditional philosophers of science, scientists are investigators trained to apply physical and conceptual tools to unravel the puzzles of nature, and whatever experiences they have as that work unfolds reflect only the subjective responses of individuals and are irrelevant to the practice of physics. But it requires at least the appetite—the curiosity, and perhaps compulsion or sense of urgency—to apply physical and conceptual tools to a puzzle in the first place, and involves gratification at its solution, or disappointment or frustration at one's inability to reach that. Science is an experientially driven activity, and thus affectively charged, its practice never without dispositions.

These dispositions existed all the way through the path to the parity violation discovery, from the initial consternation over the similarities and differences between the tau and the theta, to the famously intense discussions between T D Lee and C N Yang in their office at Brookhaven National Laboratory, during which others overheard them shouting at each other about the problem. Experimenter C S Wu dropped that romantic cruise on the *Queen Elizabeth* to go to the National Bureau of Standards to stage the experiment. In the OPA such moods of curiosity and urgency are not relevant to science. The philosopher Alan Franklin, for instance, explicitly sets them aside in some gently dismissive words. He quotes C S Wu saying that she 'had to do the experiment immediately, before the rest of the physics community recognized the importance of this experiment and did it first', but her feelings of the work's urgency, Franklin says, although interesting, may not be 'decisive evidence for the crucial nature of these experiments, or for the importance of the principle involved' (Franklin 1986, pp 22–3). Philosophers of the PPA ask, 'Why aren't they?' Science almost never advances through arbitrary experimentation; one could hardly mount a crucial experiment, or any other, without detecting a puzzle that needed to be resolved, some sense how to go about it, and the motive to pursue it.

Study of these dispositions is indeed off the grid of official intelligibility in philosophy, in a territory unexplored by the OPA or IPA. These dispositions are what one may call 'black elephants'. This wonderful term crosses two familiar expressions. One is 'black swan', a name for something whose repercussions force you to throw out key theories that you took for granted, the classic example being 'all swans are white'. The other expression it crosses is 'elephant in the room', or something whose presence everyone knows but nobody seriously addresses out of fear or embarrassment. I have heard the term 'black elephant' used in connection with environmental issues such as global warming, ocean acidification, and the growing pollution of fresh water supplies, for to acknowledge these issues on the required scale and with sufficient reality would profoundly disrupt current political activity. Many people therefore simply wave our hands at such things, admitting to them without incorporating them into plans for action.

Yet black elephants are found in science as well, especially in its dispositions. These dispositions are routinely excluded from formal accounts of how science works, and could not be incorporated without profoundly disrupting these accounts. In a sense, PPA philosophers methodically hunt black elephants in science, for they reveal how scientists retain, in their practice, deep and direct connections to the lifeworld, and how these connections have been retrained. Black elephants include the rituals often held at the commissioning or shut-down of large facilities, for instance. They also include the role of trust, whose importance for scientific inquiry remains under the radar of traditional approaches—think of Kepler's and Planck's trust in extremely slight deviations of data from expected, which might easily be dismissed as experimental error.

The puzzle-solving world that physicists inhabit is, again, rather like sport, where athletes bring their all to the ebb and flow of a game. If you spot an emotionless athlete in an exciting match, you assume they're either fantastically good at hiding their moods or are simply disengaged. Similarly, if you encounter a physicist who is blasé about their work or about their setbacks and successes, you wonder how good they could really be.

Even the notoriously impassive Paul Dirac was privately moody, as revealed by his recollection of the time he realized the likely relevance of Poisson brackets to quantum mechanics, didn't know enough about them, couldn't find them discussed adequately in his textbooks, found to his despair that the library was closed that Sunday, and then paced 'impatiently through the night and then the next morning' until the library opened.

A conventional view of science, however, labels these dispositions subjective and dismisses them as something in the domain of psychologists, not relevant to, say, physics. But there is indeed a 'physics world' that practitioners are caught up in. Usually, it's everyday stuff like conversing with colleagues and learning what others are up to, of hearing about new ideas, of reading journals and ordering supplies, of planning and carrying out new projects. Dispositions are an integral part of being, and participating, in this physics world, and thus to the activity itself. Thus in dealing with dispositions the PPA addresses an important dimension of science that is often ignored by the OPA—and this dimension is fundamental because it is more inclusive of the world of physics than both IPA and OPA.

Consider the discovery of x-rays. On the afternoon of 8 November 1895, the German scientist Wilhelm Röntgen was working alone in his lab at the Physical Institute of the University of Würzburg. While experimenting with a cathode ray tube—a vacuum tube containing an electron gun—he noticed that a nearby fluorescent screen was glowing. He noticed, as well, that when he turned off the cathode ray tube the fluorescent screen stopped glowing. Turning the cathode ray tube on and off, he found that the fluorescent screen glowed and stopped glowing. It was the most bizarre and unbelievable thing he had ever experienced; he told his wife he was afraid he was hallucinating. Hooked by the sight, he began playing around with what was happening, turning the cathode ray tube on and off, moving it around the room, and inserting things—paper, wood, metal—between the tube and the screen, to observe the effects. After eight weeks he had convinced himself that the phenomenon was real and sent off a scientific paper about it. Unsure what 'it' was, he referred to it as an 'x-ray'.

Röntgen was not seeking to confirm a hypothesis, and he was less trying to solve a puzzle than he was curious about what the phenomenon was in the first place; he was engaged in meaning-generating. Had he been blasé, he would not have pursued it.

Röntgen's discovery, for which a few years later he was awarded the first Nobel Prize in physics, embodies the six features of recognition mentioned above. It was embedded; he had no experience with invisible rays and their effects, which is why he was baffled. It was reinterpretive, for he had to keep revising his idea of what 'it' was that he was seeing—he had to fashion the idea of an invisible ray that could be emitted at one place and produce effects at another. It was temporal, for it took him two weeks to acquire the confidence that he had found something worth publishing in a scientific journal. That confidence came with the apprehension of an invariant— that he could dependably observe the phenomenon in different related ways—for certain instruments other than cathode ray tubes could produce it, screens with certain other compounds on them could glow, and set-ups could be varied in in ways with foreseeable results. He had to learn how to showcase his discovery, in the form of the photographs that he included in his journal article. The discovery had unanticipated consequences; a mere three weeks after Röntgen's announcement eight-year-old Eddie McCarthy of New Hampshire became internationally famous when x-rays were used to set a broken bone in his arm.

But the discovery also reveals several other aspects of recognition that spring from what it's like to experience the generation of meaning—to live through the process. In the oxygen discovery we mainly observe, in a third person way, an *interpretive* process—the forging of new concepts, the achieving of a solution. In the x-ray discovery, we can see a *hermeneutical* process—an individual seeking to make his experience meaningful, trying to figure out whether something he experiences is a figment of his imagination or out in the world. 'Hermeneutics' is a technical term: when philosophers focus on a process from the point of view of how the assumptions progressively change they call it interpretation, while when they focus on a process also from the point of view of how the person or community progressively changes they call it hermeneutics. Hermeneutics refers to processes that humans experience, that they live through, that generate meaning for the person involved.

Sometimes, to be sure, the research topic is assigned to a person, or taken on as part of a team or collaboration—but then the desire to tackle it comes from the mentor or team leader. Sometimes the desire fosters a long and single-minded quest, such as a several-year-long quest to find the form factor in the case of colliding protons in the 1970s. At other times, as in the case of Röntgen, the desire is to understand something noticed accidentally on the spot while doing something else; there is no reason why the screen should glow only when the cathode ray tube is on. Is an anomaly sufficient to motivate a research program? Only when one cares about it. When do you care about it? When the deviation from the norm threatens meaning.

What put Röntgen in the lab in the first place? Why did he stay there, when mysterious things started happening that made him think he was hallucinating? What made him drop what else he was doing to investigate? Others besides Röntgen had seen glowing screens, but were too wrapped up in other meaningful projects. It happens all the time in the history of science: researchers who were too busy to pursue hints of what might have been a discovery, or who were so curious that they dropped everything else. In the early 1970s the physicist Leon Lederman's data showed a bump in a histogram, which a graduate student noticed and urged him to investigate. Lederman did not, hot on the trail of the proton's form factor.

Trust is another important disposition. Everyone knows that carrying out projects you have to exercise trust. But how do you know the difference between trustworthy data and experimental error? Why did some astronomers rather than others trust the discrepancy between the predicted and observed orbit of Uranus—rather than, say, attributing it to errors in observing or recording—and therefore consider it worth investing their time to produce a model for the discrepancy? Why did some astronomers rather than others trust the model's predictions enough to drop their other projects and look for the planet that was the key element of the model? Tycho Brahe's data showed a slight discrepancy between Copernicus's planetary orbits and observed positions. His data were taken with the naked eye, not telescopes. Yet Johannes Kepler trusted the slight discrepancy enough to propose a momentous difference from Copernicus, that the planets traveled in ellipses rather than circles. Researchers from the Reichsanstalt data showed slight discrepancy from predicated value. Max Planck was sufficiently disturbed that he proposed out of 'sheer desperation' the idea of the quantum.

Dispositions are the inescapable response of an embodied creature to the world. Everyone knows that scientists feel boredom, urgency, curiosity, puzzlement. But how and where? In the discovery of Neptune, the astronomers in England, France, and Germany were all involved in other important projects; what made *this* project worth dropping the others for?

An entire set of dispositions is associated with reactions to the unforeseen. These are often lumped together by the term 'surprise'; philosophers, however, discern several different ways of experiencing the unforeseen, such as shock, bewilderment, awe and so forth. Words such as 'shock' and 'bewilderment' describe phenomena that, to philosophers, have particular technical characteristics. But then physicists do the same if you think of how you use terms like 'friction', 'impulse', and 'power'. These names might be colloquial but they are not arbitrary, and are used because of their loose relation to their technical meaning. So pay less attention to the following names and more to the experiences that they point to.

Shock. Shock is the experience you have when you trust equally strongly both a set of foundational assumptions in your inheritance and a finding that solidly conflicts with it. The vast majority of physicists, it is safe to say, experienced that emotion in 1957 following C S Wu's announcement that parity was indeed violated in polarized cobalt-60 nuclei. Shock is the momentary sense that your foundational assumptions *could* be undermined; it's a physicist's emotional acknowledgment of the abyss. Is there any testimony for this affect in scientific literature? It depends on what you mean by scientific literature. It's mentioned in Yang's Nobel lecture, and it is also evident in what happened when Yang sent to Oppenheimer—on vacation in the Caribbean—a telegram communicating the discovery of Wu's results to him. Oppenheimer cabled back, 'WALKED THROUGH DOOR'.

Bewilderment. Bewilderment is a different way to experience the unexpected. It involves a conflict between a finding and your fundamental assumptions, but this time your gut goes with those assumptions. You strongly suspect that something's wrong with an experiment or finding, but you aren't entirely sure. Think of your reaction to the announcement of the supposed discovery of cold fusion (1989) or of faster-than-light neutrinos (2011).

Surprise. Surprise is being attentively and expectantly attuned towards something which then catches you off-guard and throws you back on your own experience. You accept, let's say, both the unexpected findings and the assumptions embedded in your physics practice. But in contrast to both shock and bewilderment, you presume they can nevertheless be integrated. Surprise involves 'a believing what I cannot believe' (Steinbock and Depraz 2019). Think of the reaction when two teams of experimental particle physicists announced, on 11 November 1974, that they'd measured a spike in the number of particles produced at energies of 3.1 GeV indicating the existence of a long-lived particle now known as the J/psi. Wrote the Italian physicist G Preparata, 'It was as if one found in some remote region of the Earth a human race whose life expectancy was not 70 but rather 70 000 years!' Yet nobody in the physics community doubted either the findings or quantum electrodynamics.

Awe. Awe is a deep respect for a fundamental phenomenon when it abruptly emerges strongly and directly. Tell me you weren't awed by the first photos of a black hole, or by the 2016 data demonstrating the existence of gravitational waves. Neither event was a surprise or shock in the sense of challenging fundamental assumptions. You knew those things were surely there. What was unexpected was that they appeared so magnificently so suddenly.

Amazement. Amazement is the experience of a phenomenon that puts in an unexpected appearance but then never goes away. Think of the idea that energy comes in discrete quantities. At the beginning of the 20th century, the 'quantum' was regarded as a troublesome but isolated phenomenon that might eventually disappear, but which kept turning up, like a peculiar uninvited guest who stalks you and eventually joins your inner circle of friends.

Astonishment. Astonishment is the experience of something that you did not believe was even possible—not in the perceptual cards, so to speak—and which causes you to reconfigure your experience. In 1895 when Röntgen first saw his cathode ray tube making his fluorescent screen glow he thought he was hallucinating. Only after elaborate exploration could he believe it was real. But Röntgen remained so mystified that in his publication he called it an 'x-ray'.

These reactions are black elephants; everyone knows about them but almost never talks about them as an intrinsic part of scientific practice. For PPA they are an intrinsic part of science, and discussing them is revealing for several reasons. One is that scientists don't live through time as a sequence of discrete moments. Each disposition may be momentary, but is only possible because an individual anticipates and remembers. In these experiences, physicists live time as a flow—and therefore the way physicists live time is different from the time that they measure.

Why bother to point out that such dispositions are a familiar part of physics? One reason is the light these reactions shed on the differences and similarities between ordinary activities and physics. In physics, time is a scalar quantity, a measurable sequence of discrete moments. In ordinary life we live time as a flow in which we must simultaneously anticipate and remember. When physicists experience surprise, it dramatizes the presence of ordinary time in their activities. You can find something surprising in the present only because of your assumptions (the past) and expectations (the future). At its most exciting, physics is an encounter with the potentially strange, and when the strange arrives the encounter cannot help but be emotional.

Another lesson is that practicing physics is not a series of engineering exercises in the sense of something in which you apply a method and turn the crank. Physics involves a way of being in which the practitioners are affectively caught up with the phenomena they study; otherwise why would they be studying them? There's more to the practice of physics than seeking and finding results; it doesn't cancel subjectivity but thrives on it. Practicing physics is an encounter with the potentially strange, and when the strange arrives the result is an emotional encounter as much as a scientific one. These experiences are nearly always overlooked in standard accounts of science, but every scientist knows they are there and can be found whenever they want to notice them.

Yet another set of black elephants involves the rituals of science, such as commemoration and elation. These, too, stem from moody behaviors that everyone knows about but almost never talks about as an intrinsic part of scientific practice.

Commemoration. Everyone knows that we humans commemorate what has significantly helped and changed us—and have from the beginning—and mourn its passing. But this, too, covers a range of circumstances and emotions. In the scientific context, it would feel inappropriate if a major new scientific instrument that opened up important fields of research were commissioned \or decommissioned without cere- mony; some part of the planning process, people would think, would have failed. A ceremony was held at the 35-year-old National Synchrotron Light Source (NSLS) when it was officially shut down at 4 pm on Tuesday, 30 September 2014 (figure 4.1). The central element in the machine was an electron beam which radiated beams of x-ray and UV light—photons—used for high-precision imaging and other uses. Users often competed for who would get priority of those photons. The NSLS could be called the most productive scientific instrument ever built, if measured by such things as the number of scientists it supported and the number of research papers to emerge from its work. The shut-down ceremony attracted hundreds of people who had worked there, from all over the world, who packed the control room. A monitor was set up so that people could watch as the beam was powered down for the final time. Tears were shed. Souvenirs were passed out. T-shirts were made bearing the insistent request 'SAVE THE LAST PHOTON FOR ME'.

Elation. In any major physics department in the world you are likely to find at least one champagne bottle. Empty. Why not throw it out? But no, it's special. The label is generally signed. The champagne turns out to have been ceremoniously polished off on the occasion of some important discovery. The night before the announcement at CERN of the Higgs discovery, Peter Higgs and some others consumed a bottle of champagne. After the announcement, the empty bottle was sent on a tour around the world, and ended up in London's science museum. At one point Higgs expressed regret that he had failed to sign it. He feared, it seems, that the physics community would think that he had done something inappropriate. The practice is not limited to

Figure 4.1. T-Shirt from National Synchrotron Light Source 'Last Light' ceremony, Brookhaven National Laboratory, 30 September 2014.

physics. In a Stony Brook biology lab I saw a line of dozens of bottles above a cabinet —different years, vintages, and types. Each time a graduate student published a paper, or successfully defended a dissertation, a bottle was consumed, and when empty joined the others. Nobody thinks that those empty bottles are out of place.

Do philosophers of science really ponder the meaning of T-shirts and champagne bottles? These are physical remnants—maybe the only ones—of the dispositions of the scientists involved in such events. Yang kept Oppenheimer's telegram. Others keep empty champagne bottles. These remnants of dispositions—philosophers sometimes use state-of-mind as the technical term—is not an add-on to scientific inquiry but part and parcel of it. The dispositions are (a) fully meaningful and not mere 'subjective feelings', (b) integral to scientific practice and being a scientist, and (c) visible signs of the origins of a practice which is organized and regulated in such a way that these origins are usually hidden. They are tangible markers of expressions of what it is to make a major discovery. This is not at some deep level—it is there all the time, any time we care to pay attention. These emotions are expressions of concern; Heidegger called it technically 'care', meaning the totality of our commitments and investments and engagements in the world. Care is a fundamental structure of human being; to be is to care.

For the PPA, including the being of the inquirer into science is not only phenomenologically motivated but logically so. As affective dispositions—those motivating inquiry and responding to what it encounters are linked to scientific practice—belong to that being they must be part of a comprehensive account of scientific activity. Why, indeed, is inquiry important to inquirers in the first place? The OPA and IPA might perhaps regard what the PPA is doing here as valorizing and thematizing affective dispositions of scientists, and regard these dispositions as not essentially related to scientific activity itself. This is not what is happening. Including the being of the inquirer is simply part of any complete account of what scientific activity is in the human world.

****

Nobody who has seen the photo forgets Peter Higgs's watery eyes. Captured by photographers at CERN's main lecture theater on 4 July 2012, the image shows the British theoretical physicist holding a tissue as lab bosses announce that the Higgs boson has been discovered. Higgs, who was then 83, had welled up and was about to remove his glasses to daub his face.

Did those tears express the emotion of a particularly sensitive man? Or did they indicate moods intrinsic to life as a physicist? Higgs was not alone in experiencing feelings that day. There wasn't a single mood in the room, of course. Some were celebrating the discovery after contributing to it, or were proud of the discovery despite working in another area in or outside CERN. Others may have been dismayed at having sought but failed to contribute to it, or at having had their contributions unacknowledged. These moods were all present and inseparable from the way of life of a physicist. It is just that Higgs's was more visible—and that an alert photographer caught it on camera.

References

Beller M 1999 *Quantum Dialogue: The Making of a Revolution* (Chicago, IL: University of Chicago Press)

Crease R P 2013 The Treiman effect *Phys. World* July 13

Curd M V 1980 The logic of discovery: an analysis of three approaches *Scientific Discovery, Logic, and Rationality* ed T Nickles (Dordrecht: Reidel) pp 201–19

Dawid R 2017 Special issue: a philosophical look at the discovery of the Higgs boson *Synthese* **194** 2

Dewey J 1925 *Experience and Nature* (Chicago, IL: Open Court)

Franklin A 1986 *The Neglect of Experiment* (Cambridge: Cambridge University Press)

Hall K T 2022 *Insulin—The Crooked Timber: A History From Thick Brown Muck to Wall Street Gold* (New York: Oxford University Press) (I am grateful to Robert C Scharff for this reference)

Hanson N 1958 *Patterns of Discovery an Inquiry into the Conceptual Foundations of Science* (Cambridge: Cambridge University Press)

Kemeny J 1959 *A Philosopher Looks at Science* (Princeton, NJ: Van Nostrand)

Kuhn T 1962 *The Structure of Scientific Revolutions* (Chicago, IL: University of Chicago Press)

Peirce C S 1935 *Collected Papers of Charles Sanders Peirce* vol 5 (Cambridge, MA: Harvard University Press)

Staley K W 2017 Decisions, decisions: Inductive risk and the Higgs boson ed K C Elliott and T Richards *Exploring Inductive Risk: Case Studies of Values in Science* (Oxford: Oxford University Press)

Steinbock A and Depraz N (ed) 2019 *Surprise: An Emotion?* (Berlin: Springer)

Wartofsky M W 1978 Judgment, creativity, and discovery *Scientific Discovery: Case Studies* ed T Nickles (Dordrecht: Reidel) pp 1–16

Wootton D 2016 *The Invention of Science: A New History of the Scientific Revolution* (New York: Harper Collins)

Chapter 5

Experiment

Another way to sharply highlight the difference between the three approaches is how they analyze experimentation. We can take as an example how they would treat the events involved in the discovery of charge conjugation parity (CP) violation in 1964, an episode that followed shortly after the discovery of parity violation discussed in the previous chapter.

Here's a quick background summary. In 1961—less than 4 years after the discovery of parity violation—a bubble chamber group at Brookhaven National Laboratory (BNL) interested in strange particles (a category that had included taus and thetas) created bunches of protons, had an accelerator aim them at a steel target, used electromagnetic fields to sweep away charged particles among the fragments, directed the remainder into a bubble chamber, imaged their decays, and analyzed the results. The results were all but impossible according to a fundamental principle of particle physics.

Another group also working at BNL, like everyone else baffled by the results of their colleagues, set out to study what was happening with a different experimental arrangement designed to clarify what was happening. This group created bunches of protons, had a different kind of accelerator aim them at a different kind of target, used electromagnetic fields to sweep away the charged particles, directed the remaining particles into electronic detectors, compiled the statistics, and analyzed the results. Explaining these results required theorists to abandon the principle that had made the first group's results seem impossible. That conclusion earned this second group the Nobel Prize.

How would philosophers approach his series of events?

5.1 Frame contents (OPA)

The orthodox philosophical approach examines how experimental results are used to select which theories to believe, and how to recognize a valid experimental result. In this case it would tend to look at how the second group's conclusion was justified.

This requires more background. The second group, led by the Princeton physicists Val Fitch and James Cronin, had used BNL's Alternating Gradient Synchrotron to create kaons. Kaons came in two types, K-short and K-long, or K_s^0 and K_l^0. These two types had different lifetimes, one longer than the other, hence their nicknames. They also decayed differently—the K_s^0s decayed mainly into two pions while the K_l^0s decayed mainly into three, according to a fundamental principle of particle physics known as charge-parity (CP) symmetry. Bizarrely, however, the two types could 'mix' or 'regenerate' in such a way that one could turn into another under certain conditions. The Princeton group studied only the K_l^0s, and used electronic counters to record which pions had come from the same K_l^0 decay. The group identified enough two-pion decays, thought to be impossible for K_l^0, to suggest that CP symmetry had to be abandoned.

In *The Neglect of Experiment*, the philosopher Alan Franklin examined two roles of experimentation, one 'the choice between competing theories or hypotheses or in the confirmation and support of theories or hypotheses', the other how one comes to believe an experimental finding is real rather than 'an artifact created by the experimental apparatus' (Franklin 1986, p 3). He examined these two roles in connection with the Princeton experiment, and focused on the set of subsequent experiments that confirmed the Princeton results and the failure of all explanations that retained CP symmetry to explain how these results were justified. (Franklin does, however, attribute the 'quick acceptance' of the result to the fact that the senior members 'had a good reputation' for careful and correct work, that their result was 'statistically persuasive', and to psychological factors having to do with previous developments.)

Franklin's aim is to demonstrate that the Fitch–Cronin experiment is 'an exemplar of what one might call a "convincing" experiment', for it decisively decided the case between two mutually exclusive classes of fundamental theories, one in which CP symmetry is preserved and one in which it is not (Franklin 1986, pp 90–8). The episode, he says, therefore represents neither a Kuhnian-like paradigm shift, nor a Duhem–Quine-like situation in which discordant results can be incorporated into the prevailing theory by tinkering with the theoretical web. Franklin says of the immediately preceding experimental results, by a group led by Robert Adair of Yale: 'It was an artifact: A spurious result stimulated the work of the Princeton [Adair] group' (Franklin 1986, p 81). Franklin therefore considers the Fitch–Cronin experiment and its results as independent, from a scientific point of view, from the Adair group's experiment. He is judging the Adair group's experiment from the perspective of the Fitch–Cronin group's result.

Conceptually, the notion of an artifact seems simple and straightforward. A common definition is that an artifact is an error in an image with no counterpart in reality. Another definition is that an artifact is a behavior or structure caused by the experimenter, procedures, or equipment rather than by nature. An artifact is a false signal, the antonym of 'real effect'. The concept of artifact thus presupposes the ability to draw a distinction between our doings and nature's, what belongs to the preparation and execution of the experiment on the one hand, and to the phenomenon under scrutiny on the other. The concept has to do with what appears to be of nature, but is not. It is something that, for a while, looks like something appearing *through* our equipment, but instead is something *in* it, entirely produced *by* it. But this distinction, of course, can only be drawn *after* the experimental process is concluded, not at its beginning.

To understand experiment philosophically is at the same time to understand artifacts philosophically. Science consists in establishing the difference between artifacts and signals. Am I seeing a flaw on the plate, or a new star? Is the number on that dial about the operation of the detector, or about the properties of the proton? Is the clock's reading telling me that those neutrinos are traveling faster than the speed of light, or is that a loose junction in the link between my GPS receiver and my clock? Sometimes, in this view, we initially get the difference wrong, but in due course our investigations will uncover our mistake; we need only be clever enough in our decision procedures. We need, and can, be clever enough to formulate rules to eliminate artifacts, for the distinction between what belongs to nature and what to us already exists. For instance, as the philosopher Ian Hacking points out in his discussion of microscopes, we use different techniques, strategies, and physical processes in the investigation, and if the same signal recurs in each it is unlikely to be an artifact (Hacking 1983).

Every scientist copes with artifacts. They are part of the experimental experience, and studying them reveals much about the nature of experimentation. Distinguishing between signals and artifacts may be said to be *the* experimental skill. 'The cornerstone of experimental knowledge is the ability to discriminate backgrounds: signal from noise, real effect from artifact' (Mayo 1996, p 62). Every experimental scientist also has 'experimenter's anxiety', or worry about whether theirs is a 'real' signal or not. The question of what is artifactual sometimes has far-reaching practical consequences, as in artifacts in the chromatographic methods used in DNA analysis, or in the PCR or chemical reagent methods used in Covid testing.

Franklin called the Adair group's results an artifact because, after the Cronin and Fitch results, the Adair group's results were shown to be due to a combination of factors, only *part* of which concerned CP violation. One was the rate of parity-violating $K_1^0 \rightarrow 2p$ decays. Another was various sources of regeneration, and still a third was a small statistical fluctuation (Hawkins 1967). Adair's finding had no single theoretical explanation; it was, Franklin wrote, a product of the machine. Cronin and Fitch, by contrast, had designed the experimental equipment so that the finding of a surplus number of $K_1^0 \rightarrow 2p$ was associated with a single and unambiguous theoretical explanation, thus this was a finding about nature, a signal from 'beyond' the experiment.

This OPA approach draws a sharp distinction between signals and artifacts. Artifacts are corrected mistakes, a misrecognition of something as a fact of nature. They may stimulate inquiry, or point it in a particular direction, but they are not integral to the process itself. In the OPA, artifacts contain no interesting physics. Furthermore, the distinction between signals and artifacts is therefore one that can be made only afterwards; it can be made by a scientist, or philosopher, only after the discovery is made and confirmed; it is a distinction that must appear first in what Hanson called the 'finished research report'.

5.2 Frame changes (IPA)

The instrumental philosophical approach (IPA), by contrast, looks at the inquiry leading to a discovery. Judging the finding of an artifact can only follow upon experimental determination of what 'nature' is and is not. When I once detected a deviation from Coulomb's law in my high school physics class, my teacher informed me that it was not a new discovery but artifactual; he was sure because of the experimental work that resulted in Coulomb's law, and its acceptance and incorporation into textbooks. Experimentation involves a bootstrapping in which a finding is judged on its own rather than in the light of subsequent findings or textbooks; then that finding is used to train the next generation of scientists in how to experiment carefully and avoid artifacts.

Thus there are two kinds of artifacts. The kind I had was 'idiopathic'—unique and arising from the conditions at a specific time, a one-off result from the specific experimental set-up. But there's also 'tantalizing' artifacts, which are novel things in an experiment's performance that you can't quite make out, arise from something more general, and only find out with more experimentation. My point is that one can't tell the difference in advance, only through further experimentation. One can *guess* whether a result is one or the other by considering various factors: if the result conflicts with known, well-tested results (my finding of a deviation from Coulomb's law); if the experiment is at the limit of the equipment's accuracy (which means that probabilistic effects can play a greater role); and so on. The OPA omits the problem of distinguishing the two, for it adopts the perspective of the completed inquiry. The IPA would investigate the path of inquiry: how Adair's group set up their equipment, what puzzled them about their findings, how these findings turned from a puzzle to an anomaly, how they reacted to the anomaly, why others such as the Princeton group became involved, how the Princeton group set up their equipment, and why, in the light of the Princeton group's findings, the Adair group's were judged artifactual.

Here's more background. The Adair group began its research with no particular goal. Its methodology was simple: the group was looking, Adair said, for 'interesting events'. No hypotheses were being tested, no results replicated. K particles were also called strange particles because they were, well, strange and inherently interesting; physicists had made numerous discoveries by studying how they decayed. Shortly after K particles were discovered, for instance, the way they were produced required theorists to develop a new set of rules called associated production. The K_s^0 and K_l^0 particles, too, had the extraordinary ability to mix or regenerate when their phases interfered with each other to produce a pair of other particles, analogously to the way horizontally and vertically polarized light can be mixed in different ways to produce light that corkscrews right or left. This property was unusual as one of the first quantum mechanical phenomena that made a tangible difference as to whether a particle had one identity or another. Finally, the astonishing discovery that parity, or 'P', symmetry was violated in the weak interaction (discussed above) had been made just a few years earlier and had overturned deeply held assumptions about the structure of particle behavior. The attempt to preserve the all-important symmetries in particle physics despite parity violation had led, in fact, to the idea of CP violation as a fundamental principle. If one were looking for interesting events, the K particle system was a promising hunting ground. The Adair group was particularly interested in imaging decays of K_l^0 particles, which, again, were known to decay predominantly into three pions.

But when the Adair group members examined the results, they found more two-pion decays than expected. A small number were expected due to regeneration (mixing). Such mixing is affected by the passage of particles through matter—again, analogously to the way the phase of light is affected by passage through a medium. Some regeneration might occur when the K_s^0s passed through the walls of the bubble chamber filled with liquid hydrogen, but the effect was understood to be too weak to account for the number of two-pion decays seen.

The surplus was particularly puzzling for several reasons. A first was experimental: two previous experiments had looked specifically at K_l^0 decays, one the Brookhaven experiment that had discovered the particle (Bardon *et al* 1958), and another in the Soviet Union (Neagu *et al* 1971), in the course of which over 400 K_l^0s had been produced without a single two-pion decay observed. A second reason was theoretical: assuming the decays were from the K_l^0s, a two-pion decay was impossible according to CP symmetry. The Adair group was looking for interesting events—and instead they found impossible events.

The bare fact that the group's findings didn't fit either the theory or previous experiments was not itself significant: theories can be revised, and experiments can be wrong. But not all puzzling findings are equal, and how much attention one attracts depends on the reputation of the experimenters, the quality of their equipment, the design of the experiment, and therefore the community's trust in the results—trust being another integral dimension of scientific activity that the OPA leaves out. The Adair group's finding was a *deep* puzzle, because of the network of other things bound up with it. Each laboratory instrument is tied up in a labyrinthine way with other laboratory instruments, and these with others. The earlier Brookhaven experiment was the obviously competent one that had discovered the K_l^0 in the first place, while the Russian result was said to be an order of magnitude better. Meanwhile, not all conflicts with theory are equal, either, but this was serious. CP symmetry was intimately linked with an important theorem involving the foundations of quantum mechanics called CPT; the CPT theorem plus the finding that CP symmetry was violated would require the violation of T, time reversal symmetry, which seemed impossible. Thus this finding was not just about K_l^0 particles, but about an entire array of forefront experimental practices and fundamental theoretical structures, and suggested something was fundamentally broken in at least one of these areas. The Adair group's result was, in effect, testing this entire experimental and theoretical web. Nothing was obviously wrong with it. The group's findings grow from a puzzle to an anomaly to a crisis in the Kuhnian sense.

This is how T P Swetman views it in his article 'The response to crisis—a contemporary case study' (Swetman 1971). In this IPA lens, much more important here than the distinction between artifacts and signals was the way that the Adair result puzzled the physics community, became an anomaly, led to further inquiry, and eventually to a revolutionary conclusion. Swetman portrays what happened in these events as a Kuhnian-like crisis. He quotes Kuhn as describing a crisis as events that lead to 'The proliferation of competing articulations, the willingness to try anything, the expression of explicit discontent, the recourse to philosophy and to debate over fundamentals' (Swetman 1971, p 1320). So how did people react to this crisis?

The members of the group were able experimenters, understood the theory, were using an excellent and well-tested machine, and had built and checked the detector themselves. They, like the rest of the physics community, simply believed in CP symmetry. They knew that two previous experiments had looked at K_1^0 decays without finding a single one that had decayed into two pions. Thus they were baffled by finding them in their equipment.

Their first thought was that they had done something wrong. They therefore began fishbowling their equipment. Fishbowling means examining each of the materials and techniques involved, which is the first step in puzzle-solving. To some extent, an experiment is always being fishbowled, for equipment is monitored in standard experimental quality control and its operation monitored in the standard experimental procedure called calibration. But inevitably things have to be taken for granted, and after puzzles and anomalies these are looked at more carefully. But the three principal members of the team had 'designed and built the bubble chamber, designed and built the measuring machines, wrote the programs that handled the data, designed and ran the experiments' (Adair 1990), and understood its operation well.

Still, they checked things out. One possibility was that the two-pion decays were not coming from a K_1^0 particle. But the beauty of a bubble chamber is that it is suited to studying single events in minute detail. As it is usually surrounded by a magnetic field, tracks of charged particles curve, which yields precise information about charge and momentum. A K_1^0 decay into two pions was a two-body decay with both decay particles charged, meaning the mass of the original particle could be calculated, showing that the parent particle was indeed a K_1^0. Another possibility was that the characteristics of the accelerator beam, such as its intensity, were off, causing other effects to come into play, but these were cross-checked by the rate of other decays in the chamber, as by the three pion decays. Another possibility involved regeneration. It was remotely possible that some regeneration was due to interactions between the Ks and the copper material of the bubble chamber window. But calculations, and comparison with the operation of other experiments, showed that this effect would be too small to account for the number of two-pion decays seen. Still, this was rechecked. The group rebuilt its bubble chamber, eliminating the copper window and substituting a larger and thinner window to reduce the background in preparation for a cleaner look at the anomaly. When the thinner window was installed, another Brookhaven experimenter took the bubble chamber to a more intense K_1^0 beam at Brookhaven's other, newer and more powerful, accelerator, the Alternating Gradient Synchrotron (AGS)—but this, too, found an excess of two-pion decays over what would be expected from regeneration effects, although fewer than the Adair group.

The group's feeling of consternation mounted. Adair began to scan the bubble chamber images himself, seeing if there was anything wrong with their interpretation. Colleagues remember him coming to the group's lunch table almost daily, saying 'I found another one', and he'd pull out another image of a K_1^0 decaying into two pions. Everyone was sure that one of them was not a pion, or that there was another pion literally in the picture, but nobody could find it and everyone would scratch their heads. In what he called a conceptual leap, when he decided that what was happening was not just the malfunctioning of his equipment, Adair pulled out all events that seemed to be decays of K_1^0 k to two pions, and found twice as many as there should be.

> One possibility was CP violation, but there was no way I was going to believe in CP violation. I believed that God wouldn't do such a God-damned dumb thing—still feel that way a little! [Laughs.] So I made the mistake of trying to outguess God. I came to the conclusion that this fundamental mechanism would work if there was a force that would not be seen in any other way, but which would be enough to make this regeneration (Crease 1999).

Word of the puzzle, which was now an anomaly, spread through the physics grapevine, creating a stir. Here again, what was decisive was not the fact that it went against theory and experiment—the journals are full of such findings—but rather that this finding was so tied in with so many other experimental practices and theoretical commitments; the theory was known to be sound, the experimental equipment reliable, and the experimenters trustworthy. Occasionally a finding shakes up an entire theoretical network—think of the quantum, or relativity, or even parity violation—but these are almost always preceded by hints and forebodings of drastic change. The Adair group's result was not that. No wonder that it failed, as the French philosopher Bruno Latour would put it, to recruit enough 'allies' for the finding to become a 'black box'; that is, accepted without question. Quite the opposite; it attracted other teams who wanted to look in that box themselves. Several groups set about looking at this finding using other practices and in other areas. An experiment at the University of Illinois, inspired by the puzzle, also discovered a surplus of two pions in K_1^0 decays.

It went without saying that CP symmetry could not be 'violated', for this ran against experimental data and strongly held theoretical assumptions. Finally, there was always the possibility of something unknown; say, a hitherto undetected, weak, and long-range force between the K_s^0 and the protons in the liquid hydrogen medium of the bubble chamber that could regenerate the K_1^0s. This was the only possibility the Adair group could not definitively rule out.

The Adair group's result convinced its members enough to write it up for publication, and its work convinced enough peer reviewers for the *Physical Review* to allow it to get published despite the puzzling result. In the paper, after considering and rejecting what it called 'conventional interpretations of the data', the final paper ends as follows:

In conclusion it appears to us that the results of this experiment strongly suggest the existence of an anomalous coherent production of [K_s] mesons from a [K_L] beam. However, in view of the extraordinary consequences which may be required by such a result, it is necessary to emphasize that we cannot, at this time, completely exclude the possibility or even evaluate precisely the probability that the striking character of the data results from a combination of real effects underestimated by us together with strong statistical fluctuations (Leipuner *et al* 1963).

'Anomalous coherent production'—note the Kuhnian term, though this is about a decade before Kuhn's book *Structure of Scientific Revolutions*—is a carefully backhanded locution for 'new force'. That possibility was intriguing enough to interest theorist Steven Weinberg, and he and Adair talked about collaborating on a paper exploring its possibility.

Fitch and Cronin shared an office next door to Adair. The two found the Adair group's baffling results urgent enough—they had a strong enough discontent, another dimension omitted by the OPA—to drop everything else and turn their attention to the anomaly. They happened to be well positioned to look specifically and only at $K_1^0 \rightarrow 2p$ decays, some of which would be attributable to regeneration, but which the Adair group had found far too many of to be explained that way because of limits imposed by CP symmetry. Cronin had been running an experiment at the Cosmotron using a pair of spark chambers; the experiment had not been able to turn up anything interesting and he was thrilled to stop it and move the equipment to the AGS to look at the anomaly. The AGS was able to produce a more intense K_1^0 beam than the other accelerator, and spark chambers were ideal detectors of pion decays, for they are high-resolution devices that could observe the decays apart from any matter that might cause regeneration. They could collect statistics faster, and could be made to trigger only on two-pion decays. They found that K_1^0s indeed decay to two pions, though fewer than the other group had found, a finding which together with everything else indicated CP violation.

Now the Cronin–Fitch experiment was fishbowled. What else, other than the still-not-quite-believable explanation of CP violation, might have caused the K-longs to regenerate? Was there anything that might have interacted with them? Desperate, one Soviet physicist suggested that a fly might have flown into their equipment, and become stuck in just the right spot that the incoming K-longs passed through its body, and that had regenerated them. But a quick calculation showed that the fly would have to be massive—more massive than uranium!—which was unlikely even on Long Island during the summer. A few other suggestions were made.

The physics community found the Cronin–Fitch group's result, unlike the Adair group's, almost immediately convincing. The first place at which Fitch and Cronin presented their results happened to be before an audience that included Adair and Weinberg. 'Guess we don't have to write that paper', Weinberg said.

In the IPA lens the Adair group's experiment is not a flawed experiment of which Cronin and Fitch's was the correction. The Adair group's experiment was a typical bubble chamber experiment of the day. Nothing that emerged in the subsequent analysis of that experiment suggests that it was not well executed and analyzed. Its finding, of anomalous K decay, was itself a phenomenon, a stable presence in the laboratory world, which could be and was seen in different techniques and experiments. Their finding was reliable and repeatable; it was 'real' in the sense that it could be returned to. To use a finding at the conclusion of the experimental process —the existence of CP violation—as the basis for calling the earlier finding a mistake can only be done after the inquiry is over.

Another, more aggressive, way of treating this kind of discovery is via the 'principle of symmetry' or 'strong thesis' among constructivist sociologists of knowledge, according to which adherence to all beliefs about the natural world, whether perceived to be true or false, have to be explained in the same way, with 'how nature is' having no explanatory role in the elevation of something to scientific knowledge. In experimental inquiry, we are essentially dealing only with our own categories and distinctions; what is decisive is ultimately the social processes with which we make up or change our minds. The microscope, slide, and everything on it are products of human decisions and activity, taken and executed in the service of certain interests. So, too, are the concepts and distinctions—cell, and so on—with which we recognize and understand what we see. All findings are human products. What we decide belongs to us and what to 'nature' in that environment which is entirely due to our activity is itself a function of our distinctions. So if we draw a distinction between nature and our techniques and procedures, that, too, is yet another human decision. The further and harder we look into nature with all our instruments and techniques, the more constructed what we find will be.

Bruno Latour takes this approach in *Science in Action*. There he not only explicitly refrains from making any difference between facts and artifacts in the sense of technical objects, but also refrains from making any important distinction between facts and artifacts in the sense of alleged pieces of knowledge which we have come to repudiate. All our outcomes are of one collective process of which the key elements are: 'how to convince others, how to control their behavior, how to gather sufficient resources in one place, how to have the claim or the object spread out in time and space' (Latour and Woolgar 1987, p 131). An artifact is something about which the scientific collective has changed its mind with respect to its status as knowledge. For social constructivists it is never nature but we who do the asserting. The judge has reversed their call; the prisoner is innocent not guilty, and the judge's decision is both the only ground and the ultimate court of appeal. This turns objectivity, too, into an artifact. 'The objectivity of the experimental matter of fact', Shapin and Schaffer write, is an 'artifact of certain forms of discourse and certain modes of social solidarity'; matters of fact are but 'conventions' resulting from 'negotiations between experimenters' (Shapin and Schaffer 1985). And an artifact is a function, an artifact itself so to speak, of the collective decision process. While for the objectivists, for the OPA, nature has the final say, for the constructivists nature has no say—or, rather, if it speaks, it's in our own voice, disguised; experimentation is ventriloquism.

The way I would express the difference between the IPA and the constructivist positions is that the IPA involves an interpretive process. The concepts and terms and theories we are using at the outset are transformed in and by the process of experimentation in which we employ them, in a reciprocal and circular relation from which we never escape. This sounds like an unpleasant condition, but as philosophers know it is the only condition, the hermeneutical condition, one that makes knowledge possible, and the idea that we can escape the circular process is an illusion. But—and here's the key difference from social constructivism—it is an interpretive, circular, transformative process motivated by what is actually and stubbornly encountered in the world rather than present in our heads.

As the two experiments here show, findings are not always what you expect, and may even require new terms and conclusions. Few in the experimental group or scientific community liked it in the sense of thinking that it was the whole story. One prominent experimenter even called the Cronin and Fitch experiment, in its planning stages, the 'Kill Adair' experiment, assuming that the new finding would reveal the falseness of the Adair group's finding. But there was something stubborn about Adair's finding that resisted the desire that it go away. And Adair's finding did not become an unfinding by the Cronin–Fitch result. Rather, the Cronin–Fitch result allowed a reinterpretation of the Adair result; it parsed the Adair result, showing which part was due to P violation and what part to other processes. The Adair group's work not only provided the puzzle, but pointed the way to how its solution would be sought—in parsing the role of CP symmetry.

In the OPA, the Adair group's result was incorrect, at most a spur to more experimentation. In the IPA, part of the Adair group's result was due to a phenomenon of nature that appeared more directly and openly in the Cronin–Fitch group's result. In the OPA, Adair's result did not contain interesting physics; in the IPA, it did. In the IPA, Adair's finding did not become an unfinding by the Cronin–Fitch result. Rather, the Cronin–Fitch result allowed a reinterpretation of the Adair result; it parsed the Adair result, showing which part was due to CP violation and what part to other processes.

This process is interpretive in a second sense in that the process is not arbitrary, but involves the creation of a more complex set of terms that develop, deepen, and enrich our involvements with nature in a way that makes us not want to go back, unable to go back. There will be no more parity-conserving theories—unless there are more anomalies, more crises, more findings that question our new assumptions.

5.3 Framing (PPA)

The phenomenological orientation (PPA) would look, not at the choices and decisions involved in the inquiry, but at the 'flavor' and context of the inquiry. Simply put, in inquiry some feeling of dissatisfaction with a situation or (when more explicitly amenable to articulation) some question provokes human beings to do something that may lead to an answer. The process of inquiry leads to a deepening and enriching of human engagement with the world, to what philosophers like to call an improved 'grip' on the world. To identify a puzzle is already to narrow one's focus and to adopt a specific kind of attitude; a puzzle is circumscribed and affectively charged. In the eyes of the other two approaches, the disposition of scientists that leads to inquiry is less important than the justification of the inquiry's results, or its mechanics; for the PPA, however, what's scientifically important is the being of a scientist; what it's like to live through being an inquirer into nature.

Inquiry has a quite specific, tripartite structure, whose features are characterized in terms of the 'hermeneutic circle'. One moment is the presence of a set of involvements and abilities that I already have, and which gives me my present grip on a situation. A second moment is the often vague sense—suspicion, hope, expectation—that I can acquire more of a grip; that I can get more out of this situation, that there is something to be discovered. A third moment is the presence of a sense of how to begin to get what I want from the situation given the grip I already have—how to go about struggling with the situation in order to disclose what I am seeking, even if what I eventually arrive at is different from what I originally envisioned. The point is that this process of inquiry is not an arbitrary, robotic, or stepwise affair in which one finds knowledge, applies it, then finds more, but a continuous motion in which all three moments are at work all the time. Each moment—even simple puttering around, jamming, tinkering, toying, improvising—is already a movement of interpretation, a making explicit of what I already understand, which assures, enriches, and deepens my involvements and expectations.

In the CP violation episode, for instance, the groups each had a certain grasp of experimental physics and specifically of kaon physics. They also had a suspicion that something more was to be found in that physics—Adair's group had curiosity about kaons, Fitch and Cronin's about Adair's result. Finally, they each had a sense for how to find out more about kaon physics. Each group had each of these moments in play all the time, and only because they did were they able to proceed.

The urgency of the puzzle was its affective charge, while what's circumscribed was the focus on a particular corner of the world. But a physicist does not 'encounter' the world, or even a part of it; what one encounters is a *performance*, or what had been designed, built, operated, and witnessed as part of an inquiry.

For scientific inquiry can be contrasted with other kinds of inquiry in that it is carried forward specifically by the staging of a physical happening. Experiments are first and foremost material events (except in the peculiar case of thought experiments, which test the consistency of theory and the known); it is not enough merely to think them up. Even when we seek something as apparently abstract as a number —the value of G-2 or the gravitational constant—this number is a by-product of an elaborate staging involving the preparation of some material event and its measurement. Events do not produce numbers by themselves; they do so only when the action is properly planned, prepared, and witnessed. Experiments are also meaning generating; they change knowledge. Experiments in this sense can be contrasted with demonstrations, which recapitulate already existing knowledge for a purpose: to inspire members of a small class into further investigations; to dazzle members of a large class into learning the material; to convince skeptical colleagues; to impress reporters and politicians.

These dimensions are aspects of performance. The word has a broad spectrum of meanings. Nevertheless, one important way in which it is conceived, especially in the dramatic arts, is as the conception, production, and witnessing of material events, the experience of which gives us something more than what we already have. When viewed in this way, the structure of performance is not a metaphor that is extended merely suggestively from the theater arts into experimental science, but is the same in both. (What follows is bound to sound abstract or vague to those trained to respect disciplinary boundaries, but it is no more abstract than the kind of reasoning routinely practiced by professionals throughout the arts and the sciences.) In each, the representation (theory, language, script) used to program the performance does not completely determine the outcome (product, work), but only assists in our encounter with the new. The world is wilder and richer than we can represent; what appears in performance can exceed the program used to put it together, and can even surprise and baffle us. An experiment, for instance, that has been planned and programmed on the basis of a certain theory can disclose things that cause its creators to change the theory. That, in fact, is *why* we stage performances. What, otherwise, would be the point?

Experiments do not magically beam physicists into a special world more fundamental than this one. Experiments are material events that we set in motion and then may be motivated to remake depending on how we experience. Theory, or the instructional system with which we stage the performances, shapes how we experience them. Quantum theory, for instance, is a kind of script or 'user's manual' (Fuchs 2017). It does not latch us on to some structure in the world, but frames for us our expectations of what to experience in our performances. There is nothing more fundamental than what appears in performance; what's scientifically significant is … what is there in performance. Despite our role in them, we can't experience our performances any way we wish; their behavior is stubborn. That doesn't mean there is a deeper reality 'out there', just as there is no 'deep Hamlet' of which all our Hamlets are imitations. All we have are new productions, and the lines Shakespeare wrote—I know this analogy leaves something to be desired—have to do with what to expect and how we stage the next one. In physics, as in drama, truth is in performance. In performances, the difference between subject and object has already been stripped away; the experiencer has staged the event experienced. Is that Hamlet I experience in performance some avatar of an independent one out there, or my creation? The question can only be asked after the experience of a performance. The experimenter is not standing outside in a neutral position, but always already invested.

Performances can be classed in three kinds: mechanical expectations, standardized performances, and artistic performances. Mechanical repetition is exemplified by CDs, videos, and player pianos, which are encoded with signals that cause a device to recreate a performance. Demonstrations in science museums are often like this. But the result, no matter how beautiful or striking, is not a creation, only the echo of one. No uncertainty exists about the outcome. Standardized performances (let's say my Coulomb test) go, and are expected to go, as expected. Artistic performance is a special type that coaxes something into being, something that has not previously appeared. It goes beyond the standardized program, it is action at the limit of the already controlled and understood, it involves engagement and risk. It brings something into material presence via the hermeneutical process described above.

Performance, too, cannot be thought of in terms of its product alone; perform-ances must be *produced*, that is, prepared by an advance set of behaviors and decisions. Production refers to the set of decisions made in advance of a performance necessary for it to take place at all—which standardizes the background, and makes it possible to speak of 'the same' kind of performance.

Performances have three features: they are presentational, are related to a representation, and involve recognition. They are *presentational* in the sense that they aim at being original, disclosive, and revelatory rather than imitative or echoing. Were it otherwise, performances would be superfluous; the performance of an experiment would reflect merely the theory with which it was staged. But because the world is richer than we can represent, performances exceeds what goes into them—and that excess is why we stage them. Instruments and theories that we understand well can be set up and operated to show us things that may lead us to improve the instruments or change the theories. In this sense we may speak of a 'primacy of performance': what occurs in performance orients and sustains the skills and theories that went into it.

Performances are related to a *representation* in that this law-like behavior is represented or 'programmed' in part by texts, scripts, scores, frameworks, and so forth which are then correlated with techniques and practices so that a phenomenon appears—the work, the finding. Expressed in phenomenological language, a repre-sentation read *noetically* (with respect to its creating) is something to be performed; read *noematically* (with respect to the product, the creation), it describes the object appearing in performance. A theory both tells us how to realize a phenomenon (an electron beam, for example) materially in a laboratory, and also describes the phenomenon that appears. An important complication, however, is that a scientific term (such as 'electron') can have a dual semantics, for it can refer both to an abstract term in a theory and to a physical presence in a laboratory. The difference is like that between a 'C' in a musical score and a 'C' heard in a concert hall.

Performances are staged in front of an audience suitably prepared to *recognize* phenomena in it. As per the previous chapter, recognition is a transition from ignorance to knowledge, consists of a perceptual act, is the result of a concernful engagement with the world, and has unanticipated consequences. In the last chapter I described a concept of recognition adequate to scientific inquiry; here I discuss how such a concept helps explain features of experimental activity. It helps to explain, for instance, historicity, partial recognition, and the enthralling character of successful experimental activity.

Historicity. Inquiry is a function of specific historical circumstances. Experiments, like any hermeneutical inquiry, are never from no-where at no-when. The CP violation experiments involved the equipment of the 1960s and the theory of the 1960s. They used the techniques of the 1960s to address the problems of the 1960s: K particles.

Partial recognition. A performance, conceived in this specific way, is more than the application of a *praxis*, the application of a skill or ability like carpentry or surgery; it is a *poiesis*, a bringing forth of a phenomenon, something with presence in the world, something which can be returned to and which can appear in different ways in different circumstances, thus exhibiting some law-like behavior. By 'phenomenon' I mean only something that can be returned to and recognized again and again. I know this sounds vague, and assumes sameness in difference, but that's how I mean it and how it is in real life. This definition of phenomenon merely expresses the fact that in both our ordinary living as well as in our scientific laboratories we encounter the world not as something chaotic and random, but as something having a certain structure, as ordered, as full of things we can find again.

But scientific objects, like others, are generally initially not seen with Cartesian clarity; the experimental equipment has to be adjusted to make things clear. The Cronin–Fitch experiment was set up to see only how a certain parameter—CP symmetry—functioned, which had not been the case with the Adair experiment even though CP symmetry violation 'was there'. The experimental structure had to be adjusted to make it clear. Scientific objects often have to be recognized amidst a confusing background. The process can be likened to the experience described by Merleau-Ponty of approaching what seemed to be a tangle of poles and trees on the beach and suddenly realizing that amongst them were the remains of an old, wrecked sailing ship (Merleau-Ponty 2012). At first, the spars and masts are latent and mixed confusingly with the forest bordering on the sand dune, producing a vague tension and unease, until suddenly our sight is recast and we see the ship, accompanied by a feeling of satisfaction as the tension is relieved. Scientific objects are often recognized via an analogous process. In the laboratory, however, there is an important difference, for what or useful is at first latent and then recognized in an actively structured way. We are staging what we are trying to recognize—we build both the ship and the background environment in which we try to separate it out from its surroundings. As a result, the very way we are staging it may interfere with our ability to recognize it, and we may have to alter how we stage the experiment before what we are seeking comes into relief.

Performances can be *enthralling*; they are able to exert affective power. We care about them, are invested in them; something is at stake for us in their execution and outcome. Human beings, who are predisposed to seek knowledge, find ourselves moved to inquire into things in ways that we can pursue only by planning and staging performances. Performances thus matter to us, for they affect our perceptions of nature and our place in it, in both the arts and sciences.

The thrall of performance is often described in the arts, but happens in the sciences as well. Experiments can enthrall not just students but even and especially seasoned scientists, as is clear from numerous stories from the history of science. In 1965, Lawrence Passell, a physicist at Brookhaven National Laboratory who was working at that lab's Graphite Research Reactor, was attempting to measure by a spin state of uranium, using polarized neutrons (neutrons whose spins all point in the same direction) preparation which required building an elaborate refrigerator to polarize the uranium, he and a coworker finally succeeded one night:

I remember it well. It was about 10 o'clock when we finally got the thing cooled down, and we sat there waiting to see whether we'd see any spin dependence ... The way this thing works is, you line up the nuclei in one spin orientation, then you run it for about 10 minutes with the neutrons like *so* [pointing his index finger up], then you flip the neutron spin and you run it 10 minutes *that* way [pointing down], and back and forth [alternating up and down] ... We waited 10 minutes, and realized—wow! —the count rate was down! Could we be seeing something? Maybe! So we sat there watching, and sure enough every 10 minutes we'd see this change in the count rate. And we sat there until about 2 o'clock in the morning, till we really believed that this was not just a statistical fluke, it was real. I remember I went home that, and, boy, I could hardly sleep I was so excited ... We knew we had it ... The whole thing sort of fell into to transfer this technique beyond one specific context. It was one of the great moments in my life (Crease 2003, p 270).

And Richard Taylor once recalled the impact that the first results of an experiment had on those who participated when, after years of preparation, it finally began to run in 1978. The experimenters had created polarized electrons, accelerated them, and shot them into a batch of protons and neutrons. They were looking for evidence of parity violation, which would appear in the form of a difference in the counting rate for electrons polarized in different directions:

You'd run for a while and be above the line [Taylor holds hand above an imaginary line indicating the value for unpolarized electrons], then you'd change polarization and it dropped below [plunging his hand underneath]. I remember that in just three days you went from wondering whether the experiment was going to work to being pretty sure that you knew that there was parity violation. That, I mean, *that's* why you do this business. That feeling of knowing something before anybody else. It's, ah, it's why you're here (Crease 2003, p 270).

Historians and philosophers often ignore the moodiness, the dispositions clearly evident in such stories in their accounts of science, discussing various aspects of experiments and discoveries while failing to address either the desire motivates human beings to pursue them, or the particular kind of deep satisfaction that results.

****

The orthodox perspective looks primarily at experimentation as the bringing to light of pre-existing information, and how to justify that the information is pre-existing. The instrumental perspective looks at the way experimenters bring such information to light. The phenomenological perspective looks at experimenters as one community among others, a community of inquirers, and how the way of being of that community is similar to and different from other ways of being.

References

Adair R K 1990 CP non-conservation: the early experiments *CP Violation in Particle Physics and Astrophysics* ed J Tran Thanh Van (Paris: Editions Frontières) pp 37–54

Bardon M, Landé K and Lederman L M 1958 Long-lived neutral K mesons *Ann. Phys.* **5** 156–81

Crease R 1999 *Making Physics: A Biography of Brookhaven National Laboratory 1946–1972* (Chicago, IL: Chicago University Press)

Crease R 2003 Inquiry and performance: analogies and identities between the arts and the sciences *Interdiscip. Sci. Rev.* **28** 266–72

Franklin A 1986 *The Neglect of Experiment* (Cambridge: Cambridge University Press)

Fuchs C 2017 Notwithstanding Bohr: the reasons for QBism arxiv: 1705.03483

Hacking I 1983 *Representing and Intervening: Introductory Topics in the Philosophy of Natural Science* (Cambridge: Cambridge University Press)

Hawkins C J B 1967 K_2^0 interactions, decays, and regenerative properties at 590 MeV/c in liquid hydrogen *Phys. Rev.* **156** 1444–50

Latour B and Woolgar S 1987 *Laboratory Life: The Construction of Scientific Facts* (Princeton, NJ: Princeton University Press)

Leipuner L B *et al* 1963 Anomalous regeneration of K_1^0 mesons from K_2^0 mesons *Phys. Rev.* **132** 2285–90

Mayo D 1996 *Error and the Growth of Experimental Knowledge* (Chicago, IL: University of Chicago Press)

Merleau-Ponty M 2012 *Phenomenology of Perception* (Milton Park: Routledge) (transl. D Landes)

Neagu D *et al* 1971 Decay properties of K_2^0 mesons *Phys. Rev. Lett.* **6** 552–32

Shapin S and Schaffer S 1985 *Leviathan and the Air-Pump* (Princeton, NJ: Princeton University Press)

Swetman T P 1971 The response to crisis—a contemporary case study *Am. J. Phys.* **39** 1320

IOP Publishing

Philosophy of Physics
A new introduction
Robert P Crease

Chapter 6

Theory and theoretical objects

> The ordinary layman judges the birth of physical theories as the child the appearance of the chick [from an egg]. He believes that this fairy whom he calls by the name of science has touched with his magic wand the forehead of a man of genius and that the theory immediately appeared alive and complete, like Pallas Athena emerging fully armed from the forehead of Zeus (Duhem 1955, pp 221–2).

'Theory' has multiple meanings, such as speculation, hypothesis, explanation, interpretation, confirmed mathematical package. In mathematical physics alone, theories are highly predictive and can refer to: a comprehensive account of a single area of physics, as 'quantum theory'; to a subset, as quark theory; or to structures that model effects without identifying causes, as 'effective field theories' of things like Bardeen–Cooper–Schrieffer (BCS) theories of superconductivity. In other fields, such as evolutionary biology, theories—such as the 'theory of evolution'—are open-ended, organizing existing knowledge and outlining where gaps can be filled in without making specific predictions.

Wootton points out that the word 'theory' was historically important because it had the 'useful ambiguity' of being able to refer either to 'an established truth' or to 'a viable hypothesis', thereby 'fudging the differences between those who wanted to claim indisputable truth and those who wanted to make tentative knowledge claims' (Wootton 2016, p 397). This liberated scientific activity from having to claim true knowledge. 'Thus knowledge, in so far as we have it, is not absolute but progressive, not definitive but provisional. We make progress, but unlike those who go hawking and hunting, we may never catch our prey.' (Wootton 2016, p 398)

Far from being weaknesses, these ambiguities are strengths, and why theories have been and continue to be of paramount importance in science. Furthermore, as Duhem's analogy suggests, theories have multiple philosophical dimensions. They can be studied —like the hatched chicken—for their structure and relationship to the world they've just entered, for their emergence, or for the nature of the theoretical attitude itself.

doi:10.1088/978-0-7503-2636-0ch6

6.1 Frame contents (OPA)

The OPA focuses on the hatched chicken—on a theory as complete and coherent. A classic, three-part definition is given by the philosopher Ernest Nagel: (i) an 'abstract calculus that is the logical skeleton of the explanatory system'; (ii) 'a set of rules that in effect assign an empirical content to the abstract calculus by relating it to the concrete materials of observation and experiment'; and (iii) an interpretation of the abstract calculus that relates it to 'more or less familiar conceptual or visualizable materials' (Nagel 1961, p 90).

An essential feature of theory according to the OPA is that it comes backed by evidence. As Franklin says, in a statement clearly meant to include acceptance of theories, 'Although I know of no episodes in the history of science in which scientific decisions have gone against the weight of evidence, I think that scientists in that case would have been unreasonable' (Franklin 1990, p 2).

A few black elephants turn up in such a perspective, in the form of theories that are deemed scientific, and considered robust, but lacking evidence. One is string theory. An easy OPA fix here might seem to be that string theory is 'not yet a theory'—that it is yet unhatched. But that solution is unsatisfactory inasmuch as string theorizing is nevertheless deemed scientific and been that way for half a century. Dawid's *String Theory and the Scientific Method* grasps this black elephant by the tusks, arguing that physicists can be on solid ground when accepting a theory for reasons other than testability. The book proposes three amendments to traditional arguments for acceptable theory, and in the style of OPA gives them acronyms: there may be no alternatives to the theory (No Alternatives Argument (NAA)), the theory may bring unexpected coherence or clarity (Unexpected Coherence Argument (UCA)), and its research program may be analogous to others that have succeeded (the Meta-Inductive Argument (MIA)). These motives might more colloquially be called desperation, simplicity, and emulation.

Dawid's approach can be called a 'Modified Orthodox Response' (MOR) that does whatever tinkering with theory acceptance is needed to keep the orthodox approach from getting blatantly out of touch with actual scientific practice. The MOR seeks a formal structure that appears to be independent of researchers and to govern genuine scientific practice, although it ultimately cuts corners by drawing its modifications from the practice it seeks to legislate. Still, it responds to the ambition of the OPA by seeking the structure that fits the scientific environment—the structure of the hatched chick.

6.2 Frame changes (IPA)

Theories are understood as tools in the IPA. '*Theories thus become instruments, not answers to enigmas, in which we can rest*', writes James:

> We don't lie back upon them, we move forward, and, on occasion, make nature over again by their aid. Pragmatism unstiffens all our theories, limbers them up and sets each one at work. Being nothing essentially new, it harmonizes with many ancient philosophic tendencies. It agrees

with nominalism for instance, in always appealing to particulars; with utilitarianism in emphasizing practical aspects; with positivism in its disdain for verbal solutions, useless questions and metaphysical abstractions (James 1907).

Theories can serve as tools by carrying out three roles. First, theories *organize* existing knowledge, synthesizing what researchers have learned, giving them a grip on what is possible and impossible at the moment. Electromagnetic theories, for instance, systematize present-day knowledge about electric and magnetic phenomena. Second, theories *order* such knowledge, showing how to relate what happens at one level to what happens at another, the way that electromagnetic theory allows us to relate the behavior of charged bodies to that of subatomic and astronomical phenomena. Third, theories *orient* scientists by indicating how to utilize and explore phenomena such as electromagnetism, indicating what questions need to be investigated and suggesting ways to go about investigating them. Because theories organize, order, and orient, they possess a coherent core more important than more external properties such as empirical confirmation or testability. In the eyes of the IPA, therefore, it is wrong to refuse to deem something a theory simply because, say, it collides with experimental results or may be currently untestable.

This approach is bolstered by the appearance of black elephants, in the form of clearly scientific theories that run contrary to experimental evidence the moment they are proposed—even some that later become indispensable. It happened frequently in the path to the Standard Model, for instance, when indispensable theoretical pieces predicted particles that did not exist, or did not predict particles that did. One of the most dramatic cases occurred in the emergence of the Yang–Mills theory (Crease 2016).

In 1954 C N 'Frank' Yang and Robert Mills published a proposal for a mathematical schema that might be useful for handling strong interaction. They —and anyone else acquainted with field theory—knew that their schema would not work. But over six decades Yang–Mills, as it is known, has become fundamental not only to the Standard Model of particle physics, but also to efforts to go beyond it. Yang–Mills is the loom on which modern particle physics is woven, and high-energy physics is unthinkable without it. How was it possible for such a clearly incorrect proposal to become such a seminal event in the history of physics?

In 1953 Yang went to Brookhaven for a year, where his officemate was Robert Mills, then in his last year of getting his PhD at Columbia. Yang explained his fascination with symmetry principles; that forces in nature might result from symmetries. Weyl, for instance, had attempted to explain electromagnetism as resulting from a symmetry of the phase of the wave function, with the field a counter to the symmetry. Could something similar work for the strong interaction? The two pondered the issue. Just as in electromagnetism the phase of the wave function can be shifted arbitrarily in space and time because the interaction with the electromagnetic field will cancel out the effect of the alteration, so Yang and Mills proposed to do the same for isotopic spin, hypothesizing the existence of a 'B field' to counteract the change.

Early in 1954 Robert Oppenheimer invited Yang to present the work at the Institute for Advanced Study seminar. Wolfgang Pauli was present. He had been pursuing the same line of thinking, but had quit after encountering what seemed to be a show-stopping issue—in such theories the mass of such a field had to be zero. In quantum electrodynamics (QED), a so-called 'Abelian' theory—as Pauli knew—it is fine that the force-carrying particle (photon) is zero, but extending field theory to hadrons required a non-Abelian theory in which nature requires that the force-carrying particles be massive.

The cranky perfectionist Pauli therefore interrupted Yang's presentation demanding to know the mass of the B field. Yang said he didn't know, and resumed the presentation. Pauli cut him off again with the same demand, to which Yang responded that he and Mills had reached 'no definite conclusions'. Pauli remarked, 'That is not sufficient excuse' in such a hostile way that Yang, distressed and uncertain, sat down. An awkward silence ensued, with the seminar effectively at a halt. Oppenheimer then said, 'We should let Frank proceed.' He did, but with the rest of the presentation having an awkward flavor in the shadow of Pauli's unanswered, and obviously all-important, question.

Pauli was channeling the voice of the quantum field theory of the day, and his question was on the money. The theory required massless force-carrying particles, and nature said that this was not true in the domain where they were trying to apply it. The Yang–Mills theory was therefore clearly wrong.

When Yang returned to Brookhaven, he and Mills decided to publish their work anyway. In their paper, which appeared in *Physical Review* later that year, they wrestled with the nature of the B quantum in the final section, which Yang later wrote was 'more difficult to write than all earlier sections'. In regard to its mass, the authors write, 'we do not have a satisfactory answer'. Small wonder that their work was initially regarded as only a mathematical curiosity, or what one physicist called 'recreational mathematics'.

The phrase 'Yang–Mills' can be used in two ways: to refer to the specific model proposed by Yang and Mills in 1954, and as shorthand for any non-Abelian gauge theory of the sort that is now fundamental to the standard model. The Pauli snag had to be overcome for the one to grow into the other. How it did is a long, complicated story, a saga with many plots and sub-plots; in electroweak theory these include spontaneous symmetry breaking, while in the theory of the strong interaction asymptotic freedom. The solution being that the world turns out to be different from the one about which Pauli thought he was asking. But the end result was the creation of that loom on which modern gauge field theory is woven. The most remarkable aspect of the story, to paraphrase O'Raifeartaigh (1997), is not that it took so much time, but that it came together at all.

Some discoveries not only contribute *to* science but can also tell us *about* science, and the genesis of Yang–Mills is one. How was it possible for a theory to be not yet true of the world? When it became true can't be pinned to any specific date between 1954 and, say, 1975. It required a shift in our ideas about the world, a shift that was gradual and to which the Yang–Mills proposal itself contributed and even made possible. In this way the Yang–Mills story points to a deeper understanding of theory-making in which a theory sometimes does not have to be re-interpreted to suit the world, but the world has to be re-interpreted so that a theory can fit. Theory-making is not always a matter of seeking something provable and applicable to the world, but of articulating a sense of the world that has not yet fully taken shape. In at least some cases, theory-making involves summarizing and organizing some pre-existing sense of the world that is not yet explicitly stated, before any proof or evidence (afterwards, of course, we can say it was this way all along). The 1954 Yang–Mills theory laid out what would make it possible for the resources of quantum field theory to apply. It illustrates that some perfectly respectable theories can be open to being filled in (the theory of evolution is another, though somewhat different, illustration) rather than matching the world in every detail.

6.3 Framing (PPA)

The foundations for the phenomenology of the correlation between framing and what appears in the frame of the scientific workshop—and how to order such—is contained in Husserl's book, *The Crisis of European Sciences*. The mathematical objects of science are not eidetic forms of the real, Husserl argues, but comprise a world of imperceptible exact ideal objects or unobservable postulates.

Scientists—and Husserl was thinking of natural scientists and particularly physicists—use ideal mathematical objects as their basic conceptual tools for their descriptions of the world. These ideal mathematical objects comprise a world of their own. Husserl then outlines what amounts to a distinction between laws and theories (Hardy 2013, p 34). Laws are expressed in a mathematical formalism and concern relations between ideal objects; examples are $P \propto 1/V$, $G = m_1 m_2 / r^2$, $\Delta x \Delta p \geq h/4\pi$. By themselves, these equations don't 'say' anything about the world, but merely state ideal relationships in an ideal world. When I drop a ball or squeeze a balloon, it does not affect these formulae. Using these ideal laws, Husserl wrote, scientists cloak the experienced world in 'a well-fitting garb of ideas'. The 'clothing' is mathematical formalism, the 'clothed' the phenomena being described, and Husserl found an absolute separation between the two. The distinction between laws (formalisms) and theories is why one can say 'Shut up and calculate!' while disagreeing about what one is calculating about. Theories connect the ideal relationships with the world: P with pressure, m with masses, and so on. Theories are interpretations of these formula; they do attempt to say something about the world we experience—experience understood as including what we encounter mediated by instruments in experimentation.

Hardy writes that Kant had critiqued traditional metaphysics for taking 'the subjective conditions of the systematic empirical knowledge of objects as if they were themselves objects' (Hardy 2013), in the process turning things like the ideal of reason and fully present reality into transcendental illusions. In a similar manner, Husserl criticized Galilean science for inspiring a similar transcendental illusion by assuming that its ideal objects had real standing rather than being idealized objects that had been constituted methodologically. Husserl thus recognized from the start the activity of the subject in grasping nature through natural science.

As an example of how scientists recognize the activity of the subject—of experience—in grasping nature consider the evolution of the concept 'cell'. That has been a central concept in biology for 350 years, during which time it has undergone tremendous change, from an idea thought to explain why cork floats to a basic structural unit of living matter to a certain kind of bounded physiological unit. Throughout this development—which parallels that of many other scientific concepts—there is no evidence of a 'paradigm change', but rather of continuous incremental alteration of the concept. The history of the concept 'cell', in short, is mainly science as usual.

In some applications 'cell' may be a concept best described in scientific practice, as per OPA, as something like a purely theoretical cognitive essence that is operationalized within some formal scheme. In other applications, as per IPA, it is a tool that is continually adapted in solving new puzzles. The PPA approach looks at another dimension of scientific method: the experience that an inquirer has to have to acquire and use the concept. How, for instance, does an inquirer know when the concept is operationalized, or how to adapt the tool? Philosophers of science who study framing focus more on that pre-understanding—on what happens in the mode of being in which humans inquire into nature. That mode of being is aware that its experiences are part of an ongoing, temporal process.

The concept 'cell' was introduced into biology by Robert Hooke in his *Micrographia*, published for the newly established Royal Society in 1665. Hooke had invented a compound microscope that magnified objects thirty to a hundred times, and used it to examine various biological specimens. One was a slice of cork, which under his microscope looked as if it were composed of honeycomb-like empty structures. They reminded him of monk's quarters, so he called them 'cells'. This usage was far from the modern one. Hooke did not think all living organisms were made of cells; he only thought that he had discovered why cork was buoyant. If these structures were empty or filled with something light, it would clearly explain why cork floats.

Hooke's words make clear that he was interpreting, that he did not intend this word as capturing an essence or amounting to a formal description, something rigid that further inquiry would not elaborate or change. He was trying to point other researchers in the direction he had been following. After saying that he had searched but failed to detect greater structure in the cells, Hooke decided that he has reached the limits of what appears to him, though others with more powerful microscopes may see more. '[I]t seems very probable', Hooke writes, 'that Nature has in these passages, as well as in those of Animal bodies, very many appropriated Instruments and contrivances, whereby to bring her designs and end to pass, which 'tis not improbable, but that some diligent Observer, if help'd with better Microscopes, may in time detect' (Hooke 1665). Hooke felt his graphic description summed up what he was able to see, and anticipated that future experimenters would experience different structures with different instrumentation. He was aware that he was not looking at an 'object', but rather a phenomenon from a particular—and for his purposes, not very favorable—position that others would be able to see differently, and eventually more favorably.

We also find a similar awareness on the part of later biological researchers—that they, too, were interpreting and anticipating; that they, too, were well aware that they were only taking initial steps in a process that would eventually transform their own work; that their efforts were part of an ongoing conversation. Such researchers include Schleiden and Schwann in the late 1830s, who laid the origins for cell theory. They anticipated further transformations in the notions they were proposing, following the development of more powerful microscopes. One may also find the language of interpretation–anticipation, of retaining and awaiting, in the texts not only of biological researchers, who were well aware of rapidly improving microscope technology and thus knew that researchers who followed them by only a few years would be able to see much more detailed structures, but also in the texts of physicists, both experimental and theoretical, who were also working in an evolving instrumental context. It would be strange to expect anything different—anything other than the awareness in researchers that they are grasping the phenomena they are investigating only in one particular context, that they are not grasped fully or with complete clarity, will appear differently when addressed by different instrumentation and in different contexts, and therefore that their conceptual grasp of the phenomena involves instrumental mediation and anticipation.

As Husserl wrote (Husserl 1999, p 223), a phenomenon 'calls out to us' and 'pushes us' towards appearances not given. 'There is a constant process of anticipation, of preunderstanding.' With sufficient exploration of how a phenomenon is anticipated, 'the unfamiliar object is transformed … into a familiar object'. But the phenomenon is never fully grasped in its complete presence, horizons remain, and the most one can hope for is for a phenomenon to be given optimally in terms of the interests for which it is approached in the instrumentally mediated context in which it is given. Further, because theory and instruments are always changing, an object will always be grasped newly, leading to phases in the way it is grasped historically.

Other phenomenologists approach concepts as guides to experience, giving the primacy of the scientific process, not to concepts, but that from which the concepts arise and what they indicate about what is experienced and experiencable through instrumental mediation. Heidegger addresses this through what he calls 'formal indication'. The term refers to a phenomenon in a way that does not prejudge it, but leaves open the possibility it may appear concretely in different ways in different contexts (Heidegger 1999, p 56). Formally indicative language is a provisional use of language that calls to the attention of others something that can be experienced. The discourse is 'formal' for it describes phenomena in specific enough ways that it can be recognized, activated, and experienced by others—the phenomenon is not grasped 'as it is', only pointed to externally—yet the discourse is also 'nonbinding'. Formally indicative language calls to the attention of others the potential presence of phenomena that are able, even in need of, being interpreted—phenomena that show themselves as always already interpreted, in the sense of being realized or made present in this particular way at this particular time, with the possibility of being realized or made present in other ways.

Formally indicative language does not make something present as 'the thematic object of a straightforward and exhaustive account' (Heidegger 1999, p 13). The object is not something calculated and worked out and characterized fully in advance, not a 'cognitive essence', but something always already interpreted and anticipated. Formally indicative concepts therefore make explicit the temporality of the grasp. Such a concept is always an 'anticipatory apprehension' (Heidegger 1999, p 31), an 'anticipatory forehaving' of a unity 'which prepares a path of research in advance' (Heidegger 1999, p 41). Yet it can be a struggle to formally indicate, for the overriding temptation is to operationalize the terms, turning formal indications into categories not involved with practice and surroundings. But then inquiry becomes more about the conceptual scheme than the phenomena.

Heidegger often develops his idea of formal indication by contrasting it with 'scientific' concepts. Scientific concepts, he asserts, are generic and abstract, timeless and unchanging; their connection with human being is forgotten. Formally indicative concepts, on the other hand, point to an experienced concretion, one bound to change. Heidegger used formal indication as a way to denote ontological concepts that captured what it was to live through life; notions like care and death, not things like cells and atoms. Still, the way scientists have spoken about phenomena such as 'atoms', 'electrons', and 'cells' over hundreds of years acknowledges their ability to appear differently depending on how we are positioned with respect to them, which seems to resemble formally indicating more than theoretical representation. Scientists used these concepts to address phenomena as presented to them in their surroundings; they were responding to what was coming to them through their instruments, changed how they related to a phenomenon as it came at them differently, and took it for granted that that would happen. Formally indicated concepts allow scientists to put the weight on the phenomenon through which they are apprehended, not on the concepts themselves. It allows scientists, not to confirm only what the theory organizes for them, but to adapt the theory to what they find. That is why one experiments; to find what might be called embolisms of phenomena, areas where the theory has to be stretched to cover what is encountered.

Heidegger, however, took the OPA account of scientific concepts at its word; and indeed that account dominated philosophy of science in his time. Still, a reading of scientific discovery writings—rather than textbooks, encyclopedias, or OPA literature—shows that the way concepts and categories are used leaves open the possibility that the phenomena being addressed may appear in and be experienced concretely via different ways in different contexts to different groups of experimenters. As in the cell story, these concepts do not work out and characterize fully in advance the phenomena they address in terms of a 'cognitive essence', but as something always already interpreted and anticipated, and able to be further explored. For scientific inquiry to involve formal indication means that scientists don't have to pin something down as an 'X', once and for all of time, to talk about it and explore it. The philosophical concept of formal indication is a way of making us aware of how we talk about objects in order to focus our attention on those objects, the phenomena, rather than the conceptual scheme.

My use of the term 'formal indication' is very different from Heidegger's. He used the term in philosophical discourse, while I'm using it to focus on how concepts 'indicate' for the scientists who use them. Hooke, in this focus, is not conceptualizing, but formally indicating what 'cell' means for him. He is pulling from his experience—'cell' being, of course, a metaphor—to indicate to others what he wants them to experience. He is describing what he experiences through his equipment in a direct enough way so that others can experience it.

Scientific concepts are, if only implicitly, understood by scientists themselves as having these functions. Scientists, *in practice*, hear the language of their discourse not as about the furniture of nature, not as 'cognitive essences' that are operationalized, but as indications allowing them to pick out phenomena which become bodily present in instrumentally mediated ways with possibilities of appearing otherwise given other mediations. I have heard scientists describe what they are doing in their research—their research project and aim, as it were—as 'looking at K particles', or 'tweaking lattices', or 'thinking about it'. They have no hypotheses to test, no results to regulate. No 'method', in short. Later, when they find something, they can reconstruct what would have been a hypothesis, and how it might have been confirmed, or how it was a great puzzle that they solved.

We can return, finally, to Kaiser's discussion of Feynman diagrams to highlight the difference between these three approaches to theory. The OPA might say that the theoretical value of Feynman diagrams is their capacity to represent subatomic particle interactions. As Kaiser shows, this can only have a momentary success and for narrow purposes, for paper tools mediate physicists' experience and support neither ontological representation of, nor the claim to unmediated access to, phenomena. This is true even when the diagrams are successfully used as tools, for it is precisely the mediating traits of the diagrams that make the phenomena intelligible. The IPA might emphasize the tool-aspect of the diagrams for improving physicists' inferences, concepts, or calculations. The approach would downplay the context of justification by focusing on Kaiser's account of the way different physicists interpreted the diagrams differently and described different viable ontological roles for what these tools represented and accomplished for different ends—and that this undermines the 'what' that is supposedly being justified. The diagrams cannot justify a specific ontology when physicists do not have to agree on the ontology that they are supposedly justifying. The PPA might ask a question that both the OPA and IPA pass over: who are physicists that they can appeal to theoretic paper tools for understanding and managing phenomena. Paper tools are indeed the site of ontological formations—they change what we make of quantum phenomena—but that means that paper tools are ontology-forming activities. Kaiser's account of Feynman diagrams shows that the theoretical attitude of physicists is not a pure 'knowing-grasp' of what they are studying but a mediated grasp that refines, enriches, and intensifies their experience.

****

Whenever the young physicist and aerospace engineer Theodore von Kármán (1881–1963) visited Lake Constance—then on the western fringes of the Austro-Hungarian empire—he would head down to the lakeshore to watch the seagulls. Von Kármán understood aerodynamic theory so well that, standing on the dock, he was able to use food to guide the birds into stalling while in flight.

The episode can easily be misread to mean that the gulls are flying objects governed by aerodynamic theory, or that they consciously use their implicit knowledge of aerodynamic theory to fly, and that von Kármán was using the theory to manipulate the birds' movements. The full story is more complex. In the natural environment in which he was engaged, von Kármán appropriated aerodynamic theory to momentarily transform the gulls *into* objects so that he could play with them *as* objects. In doing so, he was not showing that the gulls in the natural landscape were most fundamentally aerodynamic-obeying objects, but that he could use the theory to treat them that way. He had the curiosity and wonder, and love of play enough to see if this were possible, and what happened.

It's not that being a theorist and a nature-lover are incompatible, or that being one is at the cost of not being the other. We need only know who we humans are to be able to be both, to be able to distinguish the two, and be aware of how and when we go from one to the other.

References

Crease R 2016 Yang–Mills for historians and philosophers *Mod. Phys. Lett.* A **31** 1630007
    Reprinted in ed L Brink and K K Phua 2016 *60 Years of Yang–Mills Gauge Field Theories: C N Yang's Contributions to Physics* (Singapore: World Scientific) pp 377–86

Duhem P 1955 *The Aim and Structure of Physical Theory* (Princeton, NJ: Princeton University Press)

Franklin A 1990 *Experiment, Right or Wrong* (Cambridge: Cambridge University Press)

Hardy L 2013 *Nature's Suit: Husserl's Phenomenological Philosophy of the Physical Sciences* (Athens, OH: Ohio University Press)

Heidegger M 1999 *Ontology—the Hermeneutics of Facticity* (Bloomington, IN: Indiana University Press) (transl. J van Buren)

Hooke 1665 *Micrographia: Or Some Physiological Descriptions of Minute Bodies Made by Magnifying Glasses. With Observations and Inquiries Thereupon* (London: The Royal Society)

Husserl E 1999 *The Essential Husserl* ed D Welton (Bloomington, IN: Indiana University Press)

James W 1907 *Pragmatism: A New Name for Some Old Ways of Thinking* (Project Gutenberg ebook), #5116 https://www.gutenberg.org/ebooks/5116

Nagel E 1961 *The Structure of Science: Problems in the Logic of Scientific Explanation* (New York: Harcourt, Brace)

O'Raifeartaigh L 1997 *The Dawning of Gauge Theory* (Princeton, NJ: Princeton University Press)

Wootton D 2016 *The Invention of Science: A New History of the Scientific Revolution* (New York: Harper Collins)

IOP Publishing

Philosophy of Physics
A new introduction
Robert P Crease

Chapter 7

Quantum mechanics

Never in the history of science has there been a theory which has had such a profound impact on human thinking as quantum mechanics, nor has there been a theory which scored such spectacular successes in the prediction of such an enormous variety of phenomena (atomic physics, solid state physics, chemistry, etc).

—Max Jammer (Wiley 1974 p v)

'It is a bad sign', wrote the late Nobel laureate in physics Steven Weinberg, 'that those physicists today who are the most comfortable with quantum mechanics do not agree with one another about what it all means' (Weinberg 2017). For philosophers that is not necessarily a *bad* sign, just a possible indication of unresolved, uninterrogated, or overlooked philosophical issues.

In quantum mechanics, more than in other corners of the 'hard' sciences, inherited concepts and assumptions about the nature of science itself collide with each other, and with other areas of human experience. These collisions give rise to the seeming paradoxes, mysteries, and absurdities that prompted Weinberg to say that it's unclear 'what it all means'.

A skeptic might claim that these collisions do not require philosophical intervention and eventually will be resolved with more experimental data or by the kind of theoretical analysis of concepts routinely practiced by scientists. Surely the collisions are ignorable or bypassable with rough and ready solutions of the sort as the Copenhagen, many-worlds, pilot-wave, or other interpretations.

But such apparent solutions come at a heavy cost, for they imply the false or illusory character of everyday experience—such as 'time'—or declare the reality of bizarre, unexperienced or unexperienceable things such as half-dead, half-living critters, instantaneous collapsing, and self-interfering stuff. These solutions can then be justified with scientistic 'gaslighting' through declaring that only scientists have a true grip on the world and that what others have is dreamlike and unreal. I don't

doi:10.1088/978-0-7503-2636-0ch7

mean to imply that quantum mechanics forces one to have to choose between two incompatible notions of 'the real' or of 'the world.' The real question, and the reason philosophers need to pay attention, is the question of why there seem to be two worlds at all, and that we have to make a decision in the first place. Two worlds is far more troubling than even two tables. Such gaslighting, again, is dangerous inasmuch as it encourages skepticism, dismissal, or scientism.

Weinberg admits that more scientific research is unlikely to dispel the mysteries, confessing, for instance, that he is not even sure what an 'elementary particle' actually is (Weinberg 2018, ch 9 and 14). Weinberg's perplexity even brought him to the threshold of questioning the frame of modern science itself: that the laws of physics 'must make no explicit reference to humans'. Weinberg says, 'We may in the end have to give up this goal, but I think not yet.'

Quantum mechanics, therefore, is an exemplary case of something that calls for critical philosophical reflection. The three approaches see dispelling the gaslighting in different ways.

7.1 Frame contents (OPA)

Respecting the difference between the context of discovery and the context of justification, philosophers of the orthodox orientation look into the frame alongside scientists, assume theories represent in some way the structure of the world as it appears there, and focus principally on the logic or ontology. Philosophers of this orientation are unconcerned with the historical path that led, say, Werner Heisenberg or Erwin Schrödinger to *develop* quantum mechanics in the first place, but are interested in the logic of the formalism that they produced and how it collides with that of classical mechanics and our expectations of what a physical theory does or what nature is.

In *Particles and Paradoxes: The Limits of Quantum Logic*, Gibbins describes this approach as follows: because we cannot picture the quantum realm—make an iconic representation of it—'therefore understanding quantum mechanics must be a matter of understanding the logic of the words and the mathematics of quantum mechanics' (Gibbins 1987, p 127). He summarizes: 'quantum mechanics baffles us because we misunderstand its logic'.

Quantum mechanics is notoriously ontologically baffling because it calls into question the idea that a pre-formed reality exists 'out there'. As nonlocality and probability have taught us, adhering to realism comes at a cost. In his book *Quantum Ontology: A Guide to the Metaphysics of Quantum Mechanics*, the Dartmouth philosopher Peter J Lewis (Lewis 2016 p 37) puts this point in the context of no-go theorems: 'Any attempt to reproduce the predictions of quantum mechanics using physical properties will result in ascribing contradictory physical properties to the system'. Viewed only within a classical frame, quantum phenomena are contradictory or mysterious.

Quantum mechanics, in this approach, is ontologically challenging. One set of OPA interpretations—including the hidden variables and pilot-wave interpretations— attempts to cope with the challenge by arguing the incompleteness of quantum mechanics; more discoveries will make its ontology less mysterious by knitting it better together. Another set of OPA interpretations seeks to meet the challenge by supplementing traditional ontology through granting status to singular aspects of the theory such as the wave function; the ontology of the quantum world is more complex than our own. In still another set of interpretations slight alterations are made to the frame to make the state of the quantum system the set of relations between the system and the reference of (impersonal) observers. The specifics of an observation then depend on the reference frame of an observer, so different observers of the same system can give different true but apparently contradictory observational reports. This involves incorporating the observer as the reference-frame creator or inhabitor, the site of the contextuality of the observation. This is a variant of the OPA because the philosopher's attention remains fully directed to the frame—observers are not people but the reference frames they inhabit—and the result is the return of a naturalistic account and of the observer-observed dualism. Carlo Rovelli writes, 'Quantum mechanics is a theory about the physical description of physical systems relative to other systems, and this is a complete description of the world' (Rovelli 2007).

The orthodox orientation evaluates interpretations for their conceptual rigor and applicability given the naturalistic attitude and the frame with which the scientists themselves are operating. The subject is on one side of the frame, the structure of nature on the other. By focusing on the collision between the structure of quantum mechanics and that of the classical frame, this first approach pinpoints exactly why we have so much trouble with its meaning; why Weinberg is so dissatisfied.

We might characterize what OPA seeks to recover as 'Same Old Universe'.

7.2 Frame changes (IPA)

The OPA also helps to understand the motivation for the instrumentalist orientation. While the OPA focuses on the logic and ontology of the world as it appears in the frame, the IPA focuses on epistemology, viewing theories as structures by which we know about the world, and new true theories as those that resolve puzzles in existing ones. The challenge of quantum mechanics does not exist 'out there' in a logic to be deciphered but 'in us' in the sense of a task to be accomplished.

This orientation thus rejects the distinction between the context of discovery and context of justification as insufficient to accurately describe inquiry. It focuses on the reasons why physicists a century ago found themselves forced to create a formalism whose meaning they still can't agree to. That is, the IPA focuses not on the contents of the frame or on the possible ontologies of what's found in it but on what puzzles we had to solve to come up with the formalism of quantum mechanics, and on judging the difference that its various interpretations make. What puzzles do these interpretations resolve? If none, why does it matter? Can a pilot-wave-er and a multiple-world-er collaborate effortlessly scientifically? In IPA inquirers are puzzle-solvers and reconstructors of the world. In its focus on inquiry offers a competing vision to that of the OPA.

The IPA thus implies a kind of ontological agnosticism but without any lack of rigor. As Stanley Fish said, again, of pragmatism, 'Pragmatism does not say that anything goes. It says that anything that can be made to go goes' (Fish 1999).

Dewey read his own pragmatism into quantum mechanics as an interpretation. He saw quantum mechanics as confirming his 'spectator theory of knowledge'—but as embodying, more prominently than other fields of science, one dimension of the pragmatist view of knowledge. Pragmatism, Dewey thought, had effectively already anticipated Bohr in Bohr's understanding of participation. In quantum mechanics, Dewey wrote, 'What is known is seen to be a product in which the act of observation plays a necessary role. Knowing is seen to be a participant in what is finally known.' Dewey continued, 'The principle of indeterminacy thus presents itself as the final step in the dislodgment of the old spectator theory of knowledge. It marks the acknowledgment, within scientific procedure itself, of the fact that knowing is one kind of interaction which goes on within the world' (Dewey 1929).

Thomas Kuhn, likewise a philosophical instrumentalist, paid little attention to the difference between context of discovery and context of justification in describing how historical events created a puzzle that forced adoption of a new ontology incommensurable with the old. The two paradigms are so discontinuous that scientists have to leap conceptually from one to the other. For Kuhn, the emblematic instance of a frame change was the one that took place in physics between classical and quantum mechanics. In such an account, the existence of paradigm changes shipwrecks interpretations tethered to ontologies. Physicists are active agents whose puzzle-solving activity reconstructs the worldly horizons.

We might characterize what the IPO ends up with as a 'Participatory Universe'.

7.3 Framing (PPA)

The phenomenological philosophical orientation proceeds, neither by trying to build or supplement ontology, nor by bypassing it and trying to resolve epistemological issues, but by interrogating our experience and its role to ask what creates the failure to arrive at a satisfactory interpretation in first place. This amounts to a rejection of science as only methodological, as consisting of agents who apply rules or reconstructions on the way to grasping an already structured world. Rather, this approach begins by trying to articulate what it is that scientists experience as being unresolvable.

Let's explore Weinberg's reaction informally as a guide, a reaction by a scientist that's not just his own subjective feeling but tied up with a shared understanding of the world.

Why, for instance, was Weinberg so disturbed that quantum physicists do not agree on 'what it all means'? A phenomenologically oriented philosopher zeros in on that last term. His use of that word indicates that he experienced frustration over, not the use, but the *meaning* of quantum mechanics. Weinberg was surely especially disturbed that the breakdown of meaning was happening in his own field. Physicists supposedly concern themselves with fundamental knowledge, but neither he nor his colleagues were seemingly able to turn it into a resolvable problem.

That experience of frustration is a phenomenological clue. Quantum mechanics forces us to face the question of the meaning of meaning. Only when we do so can we hope to supply an answer that Weinberg and his colleagues might accept. For someone to give us the meaning of something involves their relating it to what we encounter elsewhere in our experience. We learn the meaning of an institution, a book, or a word when we can connect it to what else we know. Weinberg and company have a clear grasp of quantum phenomena and its formalism, and can make practical and reliable real-world uses of it; it is scientifically coherent. But he and his colleagues cannot relate what happens in these episodes with what happens in their other experiences, and so find it mysterious; it is not experientially coherent. If they find themselves unable to articulate the relation, they can—again—ignore it, deny it, or cover it over with a conceptual band-aid that does not suffice when closely examined. But the phenomenologist seeks to understand and articulate the relation and—again—why it matters.

To a phenomenologist, then, what is of interest is what is happening when physicists do not find quantum mechanics mysterious, and when they do. When they don't, it's because they do not find the lack of ability to articulate the relation troublesome; they do not find it necessary to relate their work to their experiences outside the lab. When they do—as per Weinberg—it's because they not only recognize their lack of ability to articulate the relation, but also have a sense that that's a failing. To a phenomenologist, then, why such physicists do and don't consider quantum mechanics mysterious is the mystery to be addressed.

Weinberg and company find that none of the available interpretations articulate what they experience with their habitual ways of speaking. Given their extensive experience with the phenomenon, what is the obstacle that prevents them from breaking habit and opening up their descriptions to one they find meaningful? What prevents them from just picking one interpretation and being done with it? A phenomenologist would first want to explore what it is that they expect. What prevents the saying of something that would make Weinberg and company go, 'Aha! *That's* what it means!'

Weinberg tells us. He writes that he was looking for 'a physical theory that would allow us to deduce what happens when people make measurements from impersonal laws that apply to everything, without giving any special status to people in these laws'. What he expects is some scientifically sound narrative that links quantum phenomena to the frame which he has been using, a frame in which human beings are absent.

Weinberg is therefore treating the meaning of quantum mechanics as a *scientific* challenge, a threat to scientific coherence, whose solution will preserve the research frame. If we work hard enough, we can find some interpretation that relates quantum phenomena to the experience of the objects in the rest of the frame in a scientifically justifiable way—justifiable scientifically in the sense Weinberg has defined it.

A useful comparison here might be the theta–tau puzzle. The episode was so earth-shaking—the controversy so deep, the solution so incredible—that, as Yang said, it was like the lights suddenly switching off and being in such darkness that you were not sure you were still in the same room. And in 1956, at the 6th Rochester conference, J Robert Oppenheimer made famous Delphic remarks: 'The t meson will have either domestic or foreign complications. It will not be simple on both fronts.' No one seems to have known exactly what Oppenheimer meant, but everyone loved the remark anyway. In my interpretation, which Oppenheimer may or may not have intended, a domestic complication involves science, while a foreign complication involves philosophy. The theta–tau challenge may be scientific, philosophical, or both.

The theta–tau problem's solution turned out to be domestic. The community of physicists drew two seeming lessons. The first was scientific: that even the most apparently intractable puzzles can be solved by going behind the phenomenon to a structure of which it is a profile—that all puzzles have domestic solutions. The second lesson was moral: that physicists can blind themselves with untested assumptions. These two lessons left intact the existing subject–object structure of physics activity—the first lesson blamed the puzzle on the other side of the interaction, on what's out there, the second blamed it on this side, on our own behavior; we weren't self-aware enough. The two lessons thus reinforced naturalism, and the strength of the conviction that the resolution of any challenge would be scientific.

Weinberg intuits that quantum mechanics does not pose this kind of challenge. He sensed that it's opposite from the kind of challenge that physicists generally encounter. Usually when physicists experience a phenomenon, such as the theta–tau decay puzzle, which is meaningless in the context of everything else they know, they seek ways to tie that experience to the rest of the formalism with which they are familiar. When something seems meaningless, we add something, or remove something, and then it all fits. We add parity violation, another kind of neutrino, symmetry breaking or massive charged bosons; or we remove parity symmetry, ether, or phlogiston; and now everything fits. Solving the puzzle is a question of mechanics, not reconstruction or reinterpretation. Meaninglessness indicates the existence of a scientific mystery to be solved.

But quantum mechanics seems to involve the reverse kind of mystery. Those skilled in the formalism are unable to articulate what it means. They are unable, that is, to tie it to the rest of their experiences, experiences that they have in and out of the laboratory. Is the experience of dissatisfaction significant? Does it mean there is more scientific research to be done, as an OPA philosopher might say? Or does such an experience, or any experience, not matter because all that does is whether the mathematics work, as an IPA philosopher might think? Or does it not matter because we simply have to believe what the theory indicates like it or not? For a phenomenologist, the dissatisfaction matters because lack of fit indicates breakdown of meaning, a meaning that makes laboratory experience possible and worth conducting in the first place.

Weinberg and company seem to be hoping that some scientifically justifiable reason will emerge, a reason, say, to select the hidden variable, many-worlds, or some other interpretation. That scientific reason may be so radical that, as Yang said, when the lights come back on, it may not be the same room. But a century later, the lights have not yet come back on. And when they do, things like entanglement and the fundamental role of probability may mean not so much that it's a different room, but that it will be a different kind of light. Here the meaninglessness suggests a philosophical mystery (meaningless is a territory that not only scientists but also philosophers love). The consequences, Weinberg suspects, will indeed be foreign.

For what Weinberg wanted was a world structured by impersonal laws that are independent of humans, and he sees no way to go there. He not only sees the problem as not being able to go there, he also intuits, and is afraid, that if one is in quest of meaning there is no way *not* to go there. His frustration brought him to the threshold of questioning the frame of modern science—that the laws of physics 'must make no explicit reference to humans'.

I am describing Weinberg's reactions, not because he is a Nobel laureate, prolific explainer of physics, or somehow representative of physicists, but because his widely shared dissatisfaction with quantum mechanics is revealing. His dissatisfaction cannot be satisfied by either the OPA or the IPA; he is incapable of being comfortable with any description of stuff 'out there' independent of human thinking, or with regarding quantum theory as a mere instrument. That dissatisfaction arises, I think, from the awareness that quantum mechanics has a meaning but is incapable of being 'fit' within the frame which provides him with meaning. The source of his frustration therefore appears to lie in the framing itself. What appears in quantum mechanics is not arbitrary—it's thoroughly systematic—but cannot be revealed by the kind of framing he finds meaningful.

The phenomenological approach therefore identifies the framing as the obstacle in the way of seeing meaning in quantum mechanics, and asks what might be the meaning of the formalism without the classical frame. This is the *epoché*, or bracketing of representations, pictures, visualizables, theoretical entities, and assumptions about somethings and structures and properties and their ontologies existing out there 'behind' what appears. This doesn't mean denying the existence of those somethings and structures, only neutralizing our assumptions about them; becoming ontologically agnostic. But the *epoché* is only the first step; the second is to understand how experiencers constitute intentionalities—how they relate to their measuring experiences.

The PPA views inquiry as involving more than puzzle-solving, but as a process of anticipating—of expecting or being prepared for what might be the answer—and fulfilling (or not) these anticipations. Inquiry is therefore a temporal process in which one is aware of oneself as building on what one has experienced in the past and awaiting experiences in the future that may be different from what one experiences in the present.

One promising possibility is QBism. The term was originally short for 'quantum Bayesianism' but has lost that connection and become a self-standing term, not unlike 'CERN'. It is unlike other interpretations such as the Copenhagen, many-worlds, pilot-wave, or other interpretations because it construes laboratory measurements of quantum phenomena not as connected with beliefs about types of real or irreal objects, but as first of all actions that produce experiences, and the quantum formalism as a guide for anticipating future such experiences. 'QBism can be understood as a phenomenological reading of QM' (de la Tremblaye 2020, p 255).

The QBist perspective is phenomenological because of its strict adherence to what the researcher experiences, abstaining from assuming that these experiences are experiences of something and recognizing that such assumptions lead to the famous paradoxes and puzzles of quantum mechanics. QBism, that is, does not see the formalism as guiding beliefs in the sense of confidence that some statement is true of the world or not, but of guiding confidence in actions that produce more experiences. A QBist 'belief' is not something like a creed or assent to a description but a principle of practice. Meaningfulness lies not in the way we find stuff to fit together but in the way we interact with the world. What the formalism tells us, in the QBist perspective, is not how something out there behaves and will show itself to us the next time we measure; it shapes our expectations of what we will experience the next time we measure.

QBism is agnostic about the world independent of human thinking; it does not begin by assuming there's some structure out there that we are measuring, which then saddles us with the impossible task of reconciling other incompatible measurements with that structure given other measurements—of explaining why the 'things' found in this set of measurements mesh with those found in that set of measurements. That is precisely what produces the dissatisfaction, the mystery: Thing A and Thing B can't 'mix'. The motivation for the agnosticism is that no Structure C can be found to explain this.

Nor, however, does QBism pretend that quantum formalism is just a tool. A measurement is a new event, one in which experimenter and world are united, and it guides us to a more accurate forecast for such experiences. These rules are not subjective, for they are openly discussed, compared, and evaluated by the physics community. QBism forces moving past the language of subjectivism, for there is 'no arbitrariness in the probabilistic predictions of a quantum physicist,' no '"subjectivity" in the ordinary sense of something private' (Bitbol 2021, p 573).

In the QBist perspective, measurement is 'an action on the world by an agent that results in the *creation* of an outcome—a new experience for that agent' (Fuchs *et al* 2014). Quantum mechanics, then, 'does not deal directly with the objective world; it deals with the experiences of that objective world that belong to whatever particular agent is making use of the quantum theory', with the probabilities expressing 'her own personal degrees of belief about the event'. QBism therefore bypasses the famous paradoxes of quantum by, among other things, removing the ground on which to expect locality, for correlations between states that would look local are actually *different* experiences, made after evidence is acquired. 'A single sentence suffices for [QBists] to blow out a whole tradition of wondering about the non-local "magics" of the quantum world: "There is no nonlocality in quantum theory; there are only some nonlocal interpretations of quantum mechanics"' (Bitbol 2021, p 578). Nonlocality is resistant to traditional observation because pre-measured states cannot be known. Ignoring or failing to recognize the engagement between inquirer and what appears in inquiry also bypasses, not just quantum mechanical paradoxes associated with nonlocality, but others as well.

In this way QBism hopes to eliminate a host of traditional puzzles and paradoxes. The collapse of the wave function would become a simple updating of belief. Locality is irrelevant, because quantum correlations are not experienced by a single agent. The paradox of Wigner's friend would disappear because the friend and Wigner have different information. Even Schrödinger's cat would be rescued from existential blurring because the wave equation only describes what we ought to predict. All this by asserting that the statements of quantum mechanics 'deal only with the object–subject relation' (Fuchs *et al* 2014). All this by asserting that 'experience' is not about some property of a not-fully-given 'thing' but an interaction with the world, and 'belief' is not an assent to a thing or state of affairs but a principle of anticipation of other experiences.

In QBism measurements are understood not as disclosing a pre-existing structure, but as giving rise to personal experiences that affect how measuring subjects anticipate future measurements. One might alternately say that measurements are experiences that affects their beliefs about what they will experience in future measurements, but the term 'belief' here is not to be understood as a creed, rather in the Peircean sense as a principle of action. I may say that I don't believe in rising seas, but if I build a levee around my resort I show that I believe otherwise. For bracketing the frame does not mean that our experiences of quantum phenomena are chaotic. What is happening is not totally unconnected with us—it's not independent of our explorations—but it is not arbitrary either. We develop 'formalized anticipations' as Bitbol (2021, p 572) says; as Fuchs puts it, the quantum formalism is 'a personal accounting of what one expects' (Fuchs 2017).

QBism therefore regards physicists as always already not just 'pieces' in the world but as conjoined with them. Phenomenologists find this obvious, for physicists must form their ideas about the world the way the rest of us do: through experience. Humans are pre-connected with the world, experience comes first, and only after that experience can we ask the question of what part belongs to us and what to the world.

For phenomenologists, experience is always intentional; that is, directed towards something. Experiences are of different modes—the way I experience an emotion or a triangle or the breeze or the reading on a dial, and more, and one task of phenomenology is to describe those modes. What kinds of experiences are involved in scientific research? Experience here is not that of a novice or child in the lab but of someone trained to 'see' and 'read' instruments and measurements in a certain way. How do these experiences lead to the formalized intentionalities that anticipate future laboratory experiences? The difference between experiencing structures of the world and structures of how I experience the world is that I never mistake the latter for being permanent or timeless. Physics is an open-ended exploration that proceeds by generating ever new laboratory experiences that lead to ever more successful, but revisable, expectations of what will be encountered in the future.

Another overlap concerns the performative character of the experiments. In the QBist perspective, Adair's instrumental set-up wasn't a 'fog' that cloaked a 'reality' that Fitch and Cronin's dispelled, for each set-up showed exactly what it could. But what the latter made explicit was that researchers' expectations of future experiments would have to involve CP violation. In QBism, the subject is not an analyzer, puzzle-solver, or crafter of reference frames, but a source of experience of the ever-newly-appearing world. 'Reality is a *continuum* we partake of, not something we contemplate from without or something we encapsulate within' (Bitbol 2021, p 571). We are not spectators or digesters of the world. It makes for a situation in which 'consciousness does nothing to the physical world' (Bitbol 2021, p 570), we only stage and experience it differently.

For QBism our research explorations would not be rules that descend from heaven or from logic but are based on our experiences. What else could we base these explorations on? The experiences that experimentation provides would shape our continually revisable, more and more useful engagement with what appears. 'We rather sketch the idea of a reality whose form is continuously created as we interact with a non-specified, possibly formless, environment' (de la Tremblaye 2020, p 250). But only this provides us with 'the kind of knowledge that is needed to guide the future research of any agent, thus implying a weaker form of objectivity' (de la Tremblaye 2020, p 251). In this weaker form of objectivity we are forced to regard our anticipations of profiles as stemming, not from a stable object 'out there', but from our own expectations. We therefore can't think of our explorations as 'disturbing' anything, but only as a sampling of more profiles with which to update our expectations.

****

Let's try a thought experiment. Imagine that, around 1900, a classical explanation was somehow found for all spectra of blackbody radiation. Imagine that Planck's proposed idea of the quantum really was a mathematical trick, and that it disappeared from scientific activity as completely as phlogiston. What would be lost? From a scientific point of view, a lot. Much of the world would be impossible to understand, including light, the very stability of matter, the structure and evolution of the DNA molecule, and other such features belonging to the foundation of the world (Crease and Goldhaber 2014). But, as Bohr wrote, if quantum mechanics is a revolution 'in the history of physical science', QBism forces us to see that it is bringing about 'a revolution in our conception of knowledge' (Bitbol 2020, p 229).

References

Bitbol M 2021 Is the life-world reduction sufficient in quantum physics? *Cont. Phil. Rev.* **54** 563–80

Bitbol M 2020 A phenomenological ontology for physics: Merleau-Ponty and QBism *Phenomenological Approaches to Physics: Mapping the Field* ed P Berghofer and H A Wiltsche (Berlin: Springer)

Crease R P and Goldhaber A S 2014 *The Quantum Moment* (New York: Norton)

de la Tremblaye L 2020 QBism from a phenomenological point of view: Husserl and QBism *Phenomenological Approaches to Physics: Mapping the Field* ed P Berghofer and H A Wiltsche (Berlin: Springer)

Dewey J 1929 *The Quest for Certainty* (Carbondale, IL: Southern Illinois University Press)

Fish S 1999 Truth and toilets *The Trouble with Principle* (Cambridge, MA: Harvard University Press)

Fuchs C A, David Mermin N and Schack R 2014 An introduction to QBism with an application to the locality of quantum mechanics *Am. J. Phys.* **82** 749–54

Fuchs C A 2017 Notwithstanding Bohr, the reasons for QBism *Mind Matter* **15** 245–300 arXiv:1705.03483

Gibbins P 1987 *Particles and Paradoxes, The Limits of Quantum Logic* (Cambridge: Cambridge University Press)

Jammer M 1974 *The Philosophy of Quantum Mechanics* (New York: Wiley)

Lewis P 2016 *Quantum Ontology: A Guide to the Metaphysics of Quantum Mechanics* (New York: Oxford University Press)

Rovelli C 2007 *Quantum Gravity* (Cambridge: Cambridge University Press)

Weinberg S 2017 The trouble with quantum mechanics *New York Review of Books* 17 January http://quantum.phys.unm.edu/466-19/QuantumMechanicsWeinberg.pdf

Weinberg S 2018 *Third Thoughts* (Cambridge, MA: Harvard University Press)

Chapter 8

Magnificent structures and their foundations

Scientific 'workshops', as I've been calling the large organizational structures in which science happens, are like miniature republics. Their gross national products are scientific discoveries, and their civil servants are a network of researchers, advisors, educators, explainers, funders, publishers, technicians, and others.

From inside, these republics can seem self-sustaining, their issues and problems untethered to other areas of philosophy and human life. Seeing science from inside, too, encourages a picture of science as having no moral valence, or an intangibly beneficial one, the sort of attitude expressed in Enrico Fermi's much-cited remark that 'ignorance is never better than knowledge', or that 'science benefits humanity'. Scientism is one product of such a narrow focus, positivism another, and the dominance of STEM education yet another. Educational goals become focused on technical training and scientific advance rather than on broadly based learning and on the lifeworld in and from which these republics, and all other forms of human life, arise.

The activities inside those republics—the products and practice of scientific workshops—affect the worlds to which they belong, and the authority that they have there. For this reason philosophy of science cannot be just about scientific (or philosophical) texts. The laws, principles, and axioms of classical theory, for instance, had a huge impact in areas as diverse as art, education, religion, literature, politics, and philosophy (Dobbs and Margaret 1985, Feingold 2004). What does it mean about science that entropy shows up in the works of James Joyce, x-rays in those of Thomas Mann, and quantum imagery in John Updike's? Why does the uncertainty principle feature in literature as diverse as texts on pragmatism by John Dewey and poems by D H Lawrence? (Crease and Goldhaber 2014). What inspired an entire generation of painters to be influenced by the mathematical development of the fourth dimension? (Henderson 2013) The image of scientific activity affected not just culture but even ideas about knowledge itself. By seeming to outline criteria of objectivity, classical Newtonian theory shaped the epistemological framework of numerous philosophers, such as David Hume and Immanuel Kant. So strong was its cultural impact that Newtonian theory has been described as the foundation of modernity.

Another philosophical focus concerns the grounds from which these workshops emerged in the first place. Just as no individual is born thinking as a scientist, no culture came pre-equipped with scientific workshops. In the West, David Wootton writes, the required set of values began to develop a century or so after work such as Vesalius's on anatomy and Copernicus's on cosmology. '[A] set of values was slowly devised for how best to conduct the intellectual activity that we now call science: originality, priority, publication, and what we might call being bomb-proof: in other words, the ability to withstand hostile criticism, particularly criticism directed at matters of fact, came to be regarded as the preconditions of success. The result was a quite new type of intellectual culture: innovative, combative, competitive, but at the same time obsessed with accuracy.' Wootton continues, 'There are no *a priori* grounds for thinking that this is a good way to conduct intellectual life. It is simply a practical and effective one if your goal is the acquisition of new knowledge' (Wootton 2016, p 107).

The goals of those living in the worlds outside workshops, however, are not first and foremost those of the 'new type of intellectual culture'. The activity of workshops is not self-justifying, and are as subject to disapproval and termination from outside as any other social institution. Their existence, durability, and vitality depend on cultural endorsement and support.

Spokespeople for those inside the workshop—for scientists—who do not appreciate the difference between thinking inside this 'intellectual culture' and other varieties of thinking in the larger world can come off as seeming to be defending not just that culture, but as demanding that workshop thinking be taken as the only way that mature, civilized, and progressive people should think about everything. The motive may be educational and benign, to want to encourage appreciation of the value of science. But as I said earlier, it encourages the sense that a class of elite influencers (scientists) think they possess the truth and that the rest of the population (nonscientists) are confused and misled, making it easier for non-scientists to distrust those elites and reject their claims.

Four hundred years ago in his *Great Instauration*, Francis Bacon warned that science was in danger of becoming a 'magnificent structure without any foundation'. Humans were apt to ignore, squander, and undermine the considerable beneficial powers of scientific activity unless it was explained and justified. But Bacon also knew that securing that foundation, and confronting its vulnerability, is a different sort of activity than science itself. What it's like to think inside the intellectual culture as well as outside it, and the difference between the two, is beyond the scope of STEM education. It is yet another subject of philosophy of science, without which philosophy of science is incomplete.

***

In his fable the *New Atlantis*, written about the same time as the *Great Instauration*, Bacon addressed the problem that the existence and vitality of scientific workshops depends on the support of the world of which they are a part, and he proposed a possible way to handle the potential dangers of the gap between scientific culture and the culture of the community. How could a scientific center maintain recognition of its value by nonscientists? Much of the *New Atlantis* is a vision of Bacon's answer.

At the beginning of the *New Atlantis*, a boat of foraging European sailors—which represents humanity—is blown far off course 'in the midst of the greatest wilderness of waters in the world'. Its sailors abandon hope, prepare to die, and pray to God, but are rescued after being led to an island nation called Bensalem that is unknown to the rest of the world. Bensalem has an extensive scientific research center, called Salomon's House, and maximally benefits from its discoveries. Bacon's description of Salomon's House is an extraordinarily prophetic vision. Its laboratories are on mountaintops and deep underground; they are devoted to fields as diverse as agricultural, industrial, marine, mechanical, medical, metallurgical, optical, and pharmaceutical research; they investigate nutrition, health, and the environment; they study smells, tastes, and sounds. Salomon's House evidently provided the deliverance that the sailors unknowingly and naively took to be God's.

It is appalling that none of Bensalem or Salomon's House's leaders are women. Why couldn't Bacon, a far-sighted prophet who was familiar with female leaders in legend and reality, have foreseen a world of gender equality? The most charitable explanation is that Bacon wanted his fable to prioritize its exemplification of what organized research would look like and its value, and feared that readers would have been distracted—perhaps outraged—by a depiction of gender equality.

Still, the vision in *New Atlantis* is otherwise far-sighted and comprehensive. Besides a diversity of researchers and laboratories, Salomon's House was self-administering and had webs of publishers, administrators, teachers, planners, organizers, and explorers. Those participants were not cogs in a machine, but engaged with peers in, and morally and spiritually committed to, a common project with a specific end: 'the knowledge of Causes, and secret motions of things; and the enlarging of the bounds of Human Empire, to the effecting of all things possible', with these things being able to improve the lives of Bensalem's inhabitants.

While Salomon's House is independent and self-governing, it depends on acceptance and approval by Bensalemites, in whose midst it operates, whom it serves, and on which it depends. Bacon saw that this approval cannot be taken for granted. A special kind of 'interface' exists between the two, for the activities and interests of the communities are fundamentally different. Another key activity of Salomon's House is therefore outreach, to make Bensalem's inhabitants receptive, supportive, and recognize the value of the workshop in their midst. Museums show off and encourage Bensalemites to honor 'rare and excellent inventions', as do galleries with statues of 'all principal inventors' such as those who invented the alphabet, glass, the printing press, and various types of food. Salomon's House also cultivates Bensalemites by warning them of health and environmental endangers, along with instructions for how to cope. Daily prayers are made to God to confer blessings on Salomon's House and its quest to turn discoveries 'into good and holy uses'.

These non-scientific activities are sufficient, according to the fable, to establish Salomon's House as a stable, revered, and unchallenged presence on the island. The

Bensalemites have come to believe intuitively that any doubts about its value are infected by irrationality and ignorance. They take for granted that Salomon's House provides the eyes of their community, without which they would be in the condition of the sailors, i.e. lost in 'the midst of the greatest wilderness of waters in the world'.

Bacon's fable cannot entirely conceal one potentially disruptive and perhaps even fatal feature, exposed in a single sentence by the head of Salomon's House, in which he remarks that his colleagues make decisions about which discoveries to keep secret, for its knowledge 'we do reveal sometimes to the state and some not'. That remark glosses a severe problem. Who's the 'we' who decides, and on what grounds? Do the deciders always share the values of the Bensalemites and their leaders? Are those deciders ever wrong? Are they ever motivated by their own or by the House's self-interest? Do the researchers pursue knowledge that's abstract and done only to satisfy their curiosity, or is the knowledge always useful? Do the researchers make sure beforehand of the impact on the island of putting the discoveries to use?

Negative answers to any one of these questions would shred the trust that the Bensalemites invest in Salomon's House and in its discoveries and conclusions, and nullify everything sought by the outreach. To admit that these are even legitimate questions would reveal Salomon's House to be constructed, unstable, and not belonging to the natural order. Concealing knowledge and discoveries might insulate Salomon's House from distrust for a time, but it would amount to trying to fly beneath the Bensalemites' radar, so to speak, and if the Bensalemites ever discovered that something has been concealed from them—especially something potentially harmful—Salomon's House would have no cultural capital to respond. The foundations of its 'magnificent structures' would crumble.

The vulnerability of those foundations continues today. Well-functioning scientific equipment and even ambitious world-class laboratories worth billions of dollars have been terminated after becoming entangled in social agendas, community concerns, and political exhaustion (Crease and Bond 2022, Riordan *et al* 2015).

The *New Atlantis* offers little guidance or insight into these sorts of issues. It offers even less help in view of the fact that today's workshops do not sit, as Salomon's House did, amid a virtually homogeneous population with a shared history, experiences, and educational system in a geographically limited and insolated region. Just the opposite.

***

While the *New Atlantis* is a fable picturing the optimal and stable presence of a scientific community amid a non-scientific world, an all but diametrically opposed non-fictional episode took place in the Ottoman Empire's transition from initial hostility to acceptance of Western science. In *Learned Patriots*, the sociologist M Alper Yalçinkaya has described what happened as 'the process through which it became no longer necessary to justify the statement "The adoption of the sciences of the Europeans is the only option"' (Yalçinkaya 2015, p 210).

The Ottoman Empire once rivaled only the Habsburg and Russian empires in southeastern Europe, West Asia, and North Africa. But in the late 18th century a series of military defeats made it clear that the Empire was unable to keep up militarily with the West, thanks to scientific and technological know-how that its enemies were exploiting but that the Empire lacked. A few attempts were made to import the know-how piecemeal, such as creating an artillery school, but these failed after encountering religious, practical, and other problems. Certain Ottoman leaders became convinced that only a wider acceptance of Western science and technology would save them. But this seemed impossible given the awareness that science and technology did not 'travel alone' but required a kind of culture that was foreign to, and potentially disruptive of, Muslim culture. How could the Ottomans import something that was of foreign, and heathen, origin and still be good Muslims and Ottomans?

The measures allowing such importation involved the creation of what Yalçinkaya called an 'authority triangle'—a triangle formed by the sultan, Islam (which was equivalent to morality), and science. Each of these components added to and legitimized the authority of the other two. The ideal Ottoman citizen was one who respected and obeyed all three, and the combination of these three authorities was, in a sense, the ultimate authority that could make possible the construction of a truly deferential citizenry' (Yalçinkaya 2015, pp 217–8).

The path to reconstructing authority in the Ottoman world was through adapting the existing sources of authority of each segment of the triangle—actually, tripod. The first leg of the tripod was the authority of the Sultan; after the last Sultan in 1922 this was depersonalized and became the authority of the state. This authority was political, and charged the citizenry with acquiring the kind of knowledge that would allow the Empire to survive and prevail. In this way, respect for scientific learning went hand in hand with patriotism and good citizenship.

The second leg of the authority tripod was moral; that it was a duty to respect scientific learning. The way for this was prepared by the argument that science was ideologically and theologically neutral and not intrinsically associated with the West; not only that, but that science had been pioneered by Muslims.

The third component, which was scientific, involved the conviction that only its know-how could save the state; it had to be made 'obvious' that the Ottoman Empire needed the knowledge for self-defense (Yalçinkaya 2015, p 211).

The successful establishment of the authority tripod meant, Yalçinkaya writes, that 'the Ottoman man of science did not claim simply to be learned; he was a morally sound, reliable, and patriotic servant of the Ottoman state' (Yalçinkaya 2015, p 98). The unintended by-product of this debate, he continues, is that it required not only defending science, but at the same time that it required spelling out the nature of the community that was adopting it. As he says, 'Public discussions about science are discussions about people—people who represent, speak and act in the name of, praise, condemn, manage, fund, are exposed to, or have to somehow deal with science' (Yalçinkaya 2015, pp 14–5).

Ultimately, and very simply, Muslim Ottomans talked about people when they talked about science in the nineteenth century. The entire debate was about what kind of people the Ottomans were (and were not), and what kind of people they should (and should not) become. When Muslim Ottomans talked about science, they asked questions like 'What does familiarity with the new sciences transform a person into?' 'What does it mean to be an ignorant person?' and 'What are the virtues associated with the possession of knowledge?'... They talked about virtue and vice, laziness and industriousness, dependence and self-sufficiency, modesty and arrogance, sincerity and hypocrisy, loyalty and treachery, and contempt and deference. The meaning and boundaries of science (and for that matter, religion) were important questions to ask, but the final answers had to do with people and their qualities (Yalçinkaya 2015, pp 219–20).

The central issue in the Ottoman debate about science was thus a cultural struggle very different from the church versus state conflicts—often highly theoretical and abstract—that took place in the West. It was, Yalçinkaya remarked, 'always about what "our values" were or, even more fundamentally, who "we" were.' He continued, 'The ultimate issue was social order and the key question "Who are we and who do we want to be?"' (Yalçinkaya 2015). To a contemporary Western mind this may involve paying too much attention to the human world at the expense of the objective determinations of what the latest research in the workshop has shown us about what we should or should not do, yet this is precisely what supplies the foundations for the magnificent structures.

The Ottoman episode, as in the measures adopted in *New Atlantis*, aimed to make the citizens 'truly deferential' about science by appealing to their self-understanding rather than to some privileged epistemological status of science. Unlike the process described in the *New Atlantis*, the Ottoman process required dynamic measures tailored to the specific values of particular social groups at that one historical moment. But there is a much more important difference. While it may seem like the process involved successfully creating an interface that 'connected' two domains—the Ottoman citizenry and Western science—Yalçinkaya's remarks about the critical issue involving the citizenry's self-recognition suggest otherwise. These remarks suggest that the Ottoman citizenry came to regard science as not a tool (whatever its origin) but as intensifying and contextualizing their fundamental life experience, who they were and wanted to be. That is, they became science-*minded* rather than simply science-*using*. The successful establishment of scientific authority—'authority' of course being something one consults not surrenders to—meant first of all not acquiring scientific knowledge but acquiring insight into who one is, and *therefore* recognizing the importance of science.

***

The 'magnificent structures' of 21st century Western science have an even more fragile and uneasy relation to their foundations than those of Bensalem or the Ottoman Empire as Yalçinkaya describes it. We do not have the luxury of a solidarity arising from shared history, experiences, and ambitions. Further, we have found in our scientific structures numerous breakdowns, oppressions, bureaucracies, fragmentations, ideologies, and more reasons that provide grounds for suspicion. We cannot believe quite so quickly that scientific findings can simply be 'applied' to our lives to guide our health and welfare, and that our leaders have assumed the responsibility and commitment to do so. We have little 'civic patriotism', and no widely shared pride or excitement or deep wonder about scientific activity except for the momentary shallow wonder stimulated by such things as planetarium shows, NASA public relations photographs, popular science articles, and widely discussed but difficult to explain discussions in the scientific community such as the discovery of parity violation or the Higgs boson.

Each leg of what would be the 21st century version of the authority tripod is challengeable and challenged—and not out of ignorance, irrationality, or evil, but for sound reasons and often based on evidence. Are we serving the state by following its medicinal or scientific recommendations—or serving the interests of for-profit pharmaceutical industries and other elites and being subject to unethical clinical trials? Are we being virtuous by accepting the state's recommendations—or disempowered, compromising our self-determination? Are the vaccines safe gifts from our Sultan—or was their development a rush-job by for-profit companies or a beta-test of imperfect vaccines? Historical examples for each of these scenarios not only abound but are constantly trotted out by political forces and social media as reasons to be suspicious. Modern workshops cannot hope for deferential citizenry.

As the Ottoman episode suggests, our 'magnificent structures' cannot bootstrap their own acceptance; they cannot build their foundations using methods from their own practice; they cannot reach out by starting with ordinary theory-making and data-collecting on the inside. That would implicitly assume that there are two independent communities involved—one rational, thoughtful, scientific, and out for the good of humanity, and the other ultimately irrational and extra-scientific, motivated by greed and ignorance. Such an assumption makes crossing the divide between the two communities a matter of finding the right 'communicative resources' to be wielded by charismatic people.

Such efforts tend to make those on the outside feel guilty, patronized, or accused of being ignorant or even irrational. Such efforts can also produce the sense of scientistic gaslighting, when scientific authorities inform outsiders of concepts or findings that collide with experience and suggest that these are what's 'real', that only scientists have a true, if perhaps tentative, grip on reality, and that the experience of the rest of us reflects only illusion and ignorance. This all but guarantees skepticism or dismissal. The existence of the divide between inside and outside the workshop can contribute to the feeling of a stacked power dynamic, with those inside seeming aloof, elitist, and privileged, out of touch with what is required to survive outside. From inside the one culture, its own activity can seem vital and thriving but insufficiently recognized and respected from the outside; from the outside, it can seem to be legitimate to regard scientific activity as politicized, and therefore legitimate to pick and choose from it.

Consider the Democratic Republic of the Congo's auction of vast stretches of land for oil drilling in order to finance anti-poverty programs and to spur economic growth, an action that threatens to accelerate global warming. 'Our priority is not to save the planet', Congo's lead official in charge of climate issues told the *New York Times* (Maclean and Searcey 2022). Or consider the West Virginia treasurer's opposition to measures designed to reduce the use of fossil fuels, also quoted in the *New York Times*: 'At what cost to human flourishing are we willing to inflict these types of restrictions as it relates to access to cheap and reliable electricity?' the treasurer said. 'As West Virginians, our ability to be able to help power the nation with the natural resources that we have is a benefit not just to us, but to the entire country' (Gelles 2022).

Such reactions are only too rational, level-headed in their own way—that what's good for a sector of people or corporations in the here and now is good for everybody—and illustrate that no single tweak, fix, adjustment, communication, or mandate will make it sound to force respect for scientific authority or to ignore it. Respect for the magnificent structures is not a 'problem' to be 'solved' either by sedating or educating. It would arise from the recognition that the activity of these structures is grounded in and aims to intensify lifeworld concerns and pre-scientific experience. The biggest obstacle, strangely enough, is scientism, or the denial of the lifeworld as the ground of science. Scientific authority arises from the other direction, the recognition that science is not that from which the lifeworld is to be understood but an intensification of lifeworld activities.

Attempts to 'fix' the lack of respect for science from the inside generally take one of three postures. The 'Progressive' posture is to market the products of science and technology to the public, showing how these can benefit social agendas. But this can risk compromising their ideals and diluting their otherwise rigorous standards and practice. What might be called the 'Neo-Anabaptist' posture is for the workshop communities to keep their distance from the outside; science, in this view, belongs in the laboratory and it is up to others to use it for political ends. The 'Conservatives', finally—the vast majority—blame the schools, media, and politicians for encouraging or allowing irrationalism and pseudoscience, with the outcome great harm to humanity and the planet; the solution is to defend the workshop and promote better education as to what it's about (Hunter 2010, Crease 2011).

None of these three approaches will succeed in substantially improving the authority of science—'authority', again, meaning being something that one consults, not to which one surrenders. The principal reason for this is a failure to understand the lifeworld and how science springs from it. If untrained in social theory one can have a flawed working theory of culture and how it changes. According to this theory, a form of idealism, culture is ultimately a matter of ideas in the heads of individual actors. One can, for instance, tend to assume that one can change deeply ingrained ideas about science simply by speaking up loudly and articulately enough. This naïvely ignores how culture develops and adapts, and envisions its growth not as a matter of sense-making but as similar to the logic of science. To make science seem not just a collection of true facts, but as symbolically and culturally vital, requires creating and developing new means of exhibiting the value of scientific research in its own right, and of showing its value for addressing social problems. 'The only way to change culture is to create more of it' (Hunter 2010, p 28).

Here's a misguided picture on which to base the project of creating such a culture: Scientific findings issue from headwaters that exist below ground and emerge above ground at laboratories managed by a professional class of people (scientists) who use special kinds of methods to get these findings above ground. These headwaters are then pictured as flowing outwards into the surrounding landscape—into human life —where the findings are channeled, applied, and managed by science-using technocrats and politicians. Making science symbolically and culturally vital would therefore involve exhibiting better how these applications allow human life to flourish.

Here's another picture: dangers and threats, hopes and goals, curiosity and wonder drive humans to seek to transform their ways of life by supporting a special professional community of people (scientists) to stage and interpret performances —'above ground', as it were—based on available props, staff, and previous performances. The work of this professional community then takes on a life, and set of concerns and goals, of its own, and is sometimes tempted to tell the story that there is an underground. The relevance of the work of that special community is continually discussed, debated, dissected, circulated, evaluated, and incorporated into human life by different communities of activists, politicians, experts, journalists, leaders, and others—sometimes wisely, sometimes not; sometimes skeptically, sometimes not.

Science is thus fated to have a dual destiny, as 19th century positivists such as August Comte and John Stuart Mill understood. It is a rigorous discipline with its own norms and vitality independent of the general culture—but it is also historically and practically important to that general culture. Inside the workshops, the prevailing atmosphere encourages a sense of reality marked by the primacy of theoretical attitude and the quest for knowledge, while the activity outside is not governed entirely by the theoretical attitude and the goal 'is not the acquisition of new knowledge'.

The first step for those who would help secure the foundations of the 'magnificent structures' of science is therefore not to figure out how to communicate scientific knowledge and thinking better—still worse, to promote the idea that these are the gold standards for all knowledge and thinking—but to understand the difference between the intellectual culture of the workshops and other varieties, and become conversant with and fluent in them.

Three hundred years ago, the philosopher Giambattista Vico, well versed in both the humanities and sciences, distinguished between 'pioneers' who seek new universes and 'prospectors' who explore 'this world of ours' (Vico 1990, p 4). While pioneers attract the most attention with new discoveries and knowledge both theoretical and practical, prospectors perform a literally foundational task in exploring the ground on which the pioneers tread. Only the latter, Vico argued, makes the ground secure for the 'magnificent structures' of the former. The former need the latter because some problems that arise in the operation of the structures spring from the foundation, and the prospectors examine the coherence of the foundations—for the structures of science as well as for all other human structures. In our day, STEM prepares the pioneers, while the humanities educate the prospectors.

Today there is no reliable foundation for securing this foundation. The dominance of STEM education ironically gives rise to cracks in that foundation, and vulnerabilities in the science being practiced: in the acceptance of health-promoting measures such as vaccines, in the refusal to believe in existential threats such as global warming, and in the destruction of scientific instruments based on false information and unfounded fears.

The dominance of STEM, in short, is giving rise to neglect of the need to secure the foundations on which STEM is built. A counter is to tell stories about interactions between workshops and worlds, both disasters (Crease and Bond 2022) and positive interactions. These are not so much history lessons, but reminders of the existence, nature, and meaning of those interactions, enriching the ability to manage them in the present.

Stories about the successes and failures of safety, health, and environmental decisions, for instance, bring into the open the questions of how we want these decisions to be made in the future. Compelling and dramatic enough stories may be able to motivate rethinking of how we make those decisions: do we want them to be made the ways we have in the past? This means confronting yet another set of puzzles that scientific activity alone cannot resolve.

References

Crease R P 2011 To change the world *Phys. World* June

Crease R P and Goldhaber A S 2014 *The Quantum Moment* (New York: Norton)

Crease R and Bond P 2022 *The Leak: Politics, Activism, and Loss of Trust at Brookhaven National Laboratory* (Cambridge: MIT University Press)

Dobbs B J T and Jacob M C 1985 *Newton and the Culture of Newtonianism* (Atlantic Highlands, NJ: Humanities)

Feingold M 2004 *The Newtonian Moment: Isaac Newton and the Making of Modern Culture* (New York: Oxford University Press)

Gelles G 2022 West Virginia moves to sever dealings with banks it says don't support coal *New York Times* 29 July, A12

Henderson L D 2013 *The Fourth Dimension and Non-Euclidean Geometry in Modern Art* (Cambridge: MIT University Press)

Hunter J 2010 *To Change the World* (New York: Oxford University Press)

Maclean R and Searcey D 2022 Congo to auction land to oil companies: 'Our priority is not to save the planet' *New York Times* 25 July, A6

Riordan M, Hoddeson L and Kolb A W 2015 *Tunnel Visions: The Rise and Fall of the Superconducting Super Collider* (Chicago, IL: University of Chicago Press)

Vico G 1990 *On the Study Methods of Our Time* (Ithaca, NY: Cornell University Press) (transl. E Gianturco)

Wootton D 2016 *The Invention of Science: A New History of the Scientific Revolution* (New York: Harper Collins)

Yalçinkaya M A 2015 *Learned Patriots: Debating Science, State, and Society in the Nineteenth-Century Ottoman Empire* (Chicago, IL: University of Chicago Press)

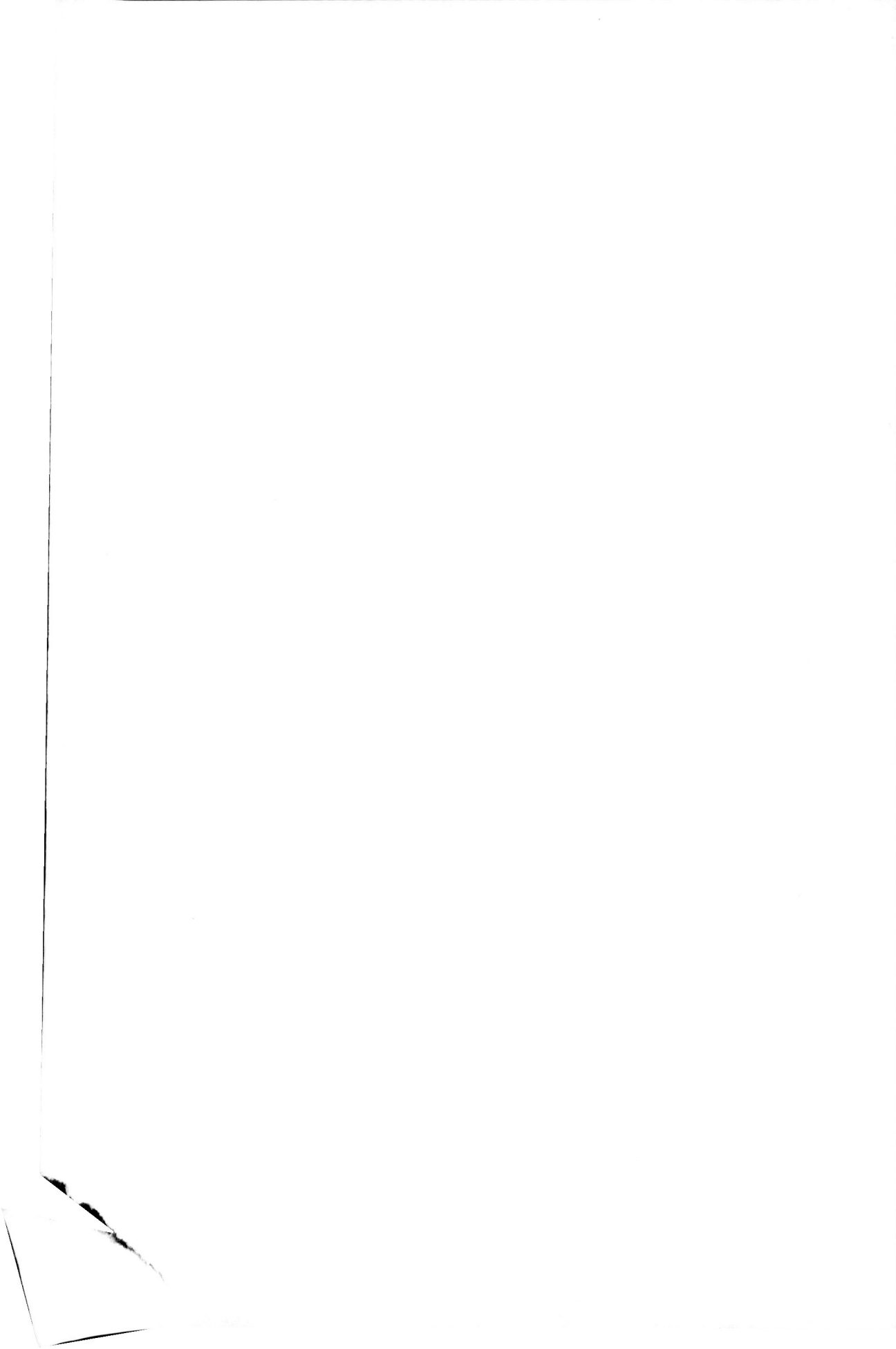

www.ingramcontent.com/pod-product-compliance
Lightning Source LLC
Chambersburg PA
CBHW050401110426
42812CB00006BA/1764